D0302961

3062788562

IMPROVING HOW UNIVERSITIES TEACH SCIENCE

Improving How Universities Teach Science

Lessons from the Science Education Initiative

CARL WIEMAN

Harvard University Press
Cambridge, Massachusetts & London, England
2017

Copyright © 2017 by the President and Fellows of Harvard College
All rights reserved
Printed in the United States of America

First printing

Library of Congress Cataloging-in-Publication Data
Names: Wieman, C. E. (Carl Edwin), author.
Title: Improving how universities teach science : lessons from the Science Education Initiative / Carl Wieman.
Description: Cambridge, Massachusetts : Harvard University Press, 2017. | Includes bibliographical references and index.
Identifiers: LCCN 2016049553 | ISBN 9780674972070 (hardcover : alk. paper)
Subjects: LCSH: Science—Study and teaching (Higher) | Science—Study and teaching (Higher)—Evaluation. | Effective teaching. | Teaching—Methodology. | Education, Higher—Aims and objectives.
Classification: LCC Q181 .W54 2017 | DDC 507.1/1—dc23
LC record available at https://lccn.loc.gov/2016049553

In recognition of Sarah Gilbert and Stephanie Chasteen, whose contributions have greatly added to the data collection, analysis, and writing of this book.

CONTENTS

IMPROVING HOW UNIVERSITIES TEACH SCIENCE

Introduction

AFTER MANY YEARS DOING RESEARCH related to improving undergraduate science education, I became convinced that it was time for broad-based change. The evidence was overwhelming that new research-based methods were superior to the lecture instruction found in most college science classrooms. It was also clear to me that the faculties of science departments were mostly unaware of this superiority, even in the situations where active research on improving science education was taking place within their own departments. Although an enormous number of individual experiments had been designed to improve single courses, none had broadened their focus to the problem of bringing the most successful teaching methods to scale. I launched the Science Education Initiative (SEI) at the University of Colorado and the University of British Columbia as an attempt to determine whether it was possible to get entire science departments to adopt these better teaching methods.

This book tells the story of the initiative. The assumption behind the decision to publish it is not that many institutions will seek to replicate the whole experimental adventure. Rather, by seeing the thinking and effort that went into it, they can be more confident that the method has produced insights they can use. Other programs attempting to improve the quality of university-level science, math, or engineering instruction can benefit from this experience. Many of the SEI's lessons learned would be valuable in the design and implementation of any size program for improving undergraduate

Table 0.1. Features of the SEI programs

	University of Colorado	University of British Columbia
Total funding	$5.3 million	$10.8 million
Funding per department	$150,000–$860,000 (avg. $650,000)	$0.3 million–$1.8 million (avg. $1.4 million)
Total number of science education specialists	24	52
Transformed courses/credit hours per year	71/53,000	164/139,000
Number of faculty who changed teaching methods (ranging from 10–90 percent of departments)	102	180

Note: Funding per department was averaged over the six fully funded departments at CU and six departments at UBC, since the three-department undergraduate biology program was funded as a single department.

teaching. Even more broadly, the information and conclusions that emerged from this effort are relevant to efforts to bring about other forms of widespread change in university settings. The SEI and the effort to compile this book yield a uniquely valuable set of observations about the operations of academic departments, about how departments can best support change, and about the many ways change efforts can fail.

It is possible to sum up the major findings of the SEI in a few headlines. First, the initiative showed that *it is possible to achieve widespread change within departments.* As Table 0.1 illustrates, this was a substantial project that had a large impact. It altered the teaching of nearly 200,000 credit hours per year at these two institutions, changing how nearly 300 science faculty went about their work in 235 courses. Major portions of faculty (up to 90 percent in the most successful departments) adopted new teaching methods, and the level of transformation (in terms of both absolute numbers and percentages of undergraduate credit hours) was substantial. There is good evidence of the sustainability of these changes, at least as measured in the short term. However, there was wide variation across the departments as to the level of success, suggesting many general lessons about what helps and hinders such educational innovation.

The SEI made clear that *virtually all faculty want to teach well, and nearly all faculty can learn to use new teaching methods effectively, but the methods recommended by the SEI do involve a significant initial learning curve.*

There are, of course, substantial challenges to implementing many kinds of change in universities. The SEI revealed that *the largest barrier to faculty change is the formal incentive system.* Faculty see the institutional incentive system as penalizing any time taken away from research to improve teaching or make use of nontraditional teaching methods. When faculty members did embrace new teaching methods, it was usually because they valued the greater personal satisfaction they would experience with students' improved engagement and learning.

In the most successful departmental change efforts, certain key elements stood out. First, *a substantial competitive grant program for departments to improve undergraduate education was clearly effective.* Second, *there was great value in having science education specialists (SESs) with expertise both in their discipline and in teaching embedded in the departments to work with the faculty.*

Third, although each department's experience differed, *the primary determinant of departmental success was the overall quality of organization and management within the department.* Each department's particular culture played a crucial role in how it viewed and carried out educational change.

Finally, it became evident that *persistence and flexibility were essential, as some of the SEI's initial assumptions were wrong and many unexpected issues arose.* Many adjustments were needed based on what was learned over the course of the SEI. These changes resulted in substantial improvements.

A Guide to the Book

Chapter 1 begins with my vision of an optimized university: one that produces the best education possible in the most efficient manner within the current resource constraints. This is the ultimate goal toward which the SEI was striving. Chapter 2 presents the model of change incorporated in the SEI, the principles behind that model, and its specific components. This is based on theories of organizational change and the adoption of innovation as mapped onto the context of a science department at a large research university, the necessary unit of change. It also incorporates my own experience at successfully transforming several courses by a specific process of backward design. Chapter 3 is a lengthy discussion of the SEI implementation. It explains how the SEI funded departments through a competitive grant process, and how departments then used the funds to support the process of changing how courses were designed and how faculty taught, assisted by SESs embedded in the departments. Changes in courses were informed by a three-pronged effort to define what students *should* be learning; to measure accurately what they *were* in fact learning; and to introduce more effective research-based instructional practices to improve that learning. Chapter 4 describes the role of SESs, whose somewhat novel position played a vital role in this innovation process. I discuss how they were hired and trained, how they typically functioned within departments, and where their subsequent career paths took them. Chapter 5 presents all that we accomplished in the SEI. Beyond the departmental-level statistics on how many courses and faculty were affected and what specific changes were made, the chapter discusses broader impacts on how these departments view and carry out educational change. Chapter 6 takes stock of the SEI's model, identifying which aspects of it worked well, which required modification, and which simply failed. In a university setting, the quality of learning hinges on faculty decisions about how to teach. This chapter offers my conclusions about what factors drive those decisions and how well the SEI was able to influence them. In the Coda, I draw together everything that I have learned from the SEI in order to advise faculty and administrators who desire to make large-scale improvements in science education at their institutions. In other words, as someone who began his work as a science educator decades ago, I share what I would have done then had I known what I know now.

ONE

The Vision

THIS IS A UNIQUE TIME in the history of science education. In recent years, those of us who have called for improvement in Science, Technology, Engineering, and Mathematics (STEM) education are receiving major attention, and there is an increasing awareness that we need to change our approach to the way we teach science. Many of these efforts, including my own, are guided by the work emerging from two rapidly growing fields: the learning sciences and, in particular, discipline-based education research (DBER) in science at the undergraduate level. DBER has produced an extensive body of research with compelling evidence that many of our current ways of teaching undergraduate science, particularly the pervasive lecture, are quite ineffective. Moreover, this new research is laying the foundation for a new model of science education by empirically testing which methods of instruction produce the best results for students. Collectively, these studies indicate that we could significantly improve the quality of science education if universities and colleges adopted these research-based methods on a large scale. Most importantly, these are changes that can be implemented today, and can be implemented within our present institutional structures and, crucially, within our current budgets. I believe, and the Science Education Initiative has gone a long ways toward demonstrating, that by adopting these new approaches to teaching, we can create a higher education system that has most of the same organizational structures and priorities—and the same price tag—as the one that currently exists, but provides far greater educational value. In this chapter, I lay out how such an improved system of higher STEM education would look.

The Educational Goal

By engaging in study within a discipline, the student should develop expertise in the subject, including problem-solving approaches and skills, habits of the mind, content knowledge, and beliefs about the nature and relevance of the subject. These learning gains should be visible both at the level of an individual course and across a curriculum or program of study as a whole. At the course level, it is important that students move toward such expert-like views of STEM—even if they are non-STEM majors taking a single course to fulfill requirements. The educational goal should be to have these students understand and think about science more in the way a scientist does, including appreciating the scientific process, relating ideas in STEM to real life, and developing curiosity about the natural world. At the program level, it is important that curricula be purposefully aligned, ensuring that courses build on one another to provide ever-deepening mastery of such core competencies.

In the modern world, there is a growing need for technical literacy and skills across the workforce and in public policy decisions.[1] This makes science education important for all students, not just those pursuing careers in science or engineering. A particularly important segment of this population for whom science education is especially important is the fraction who will become the future K-12 teachers.

There is a large and growing body of evidence indicating that post-secondary science education is failing to meet these educational needs. Although there is a particularly large amount of research on how students learn physics and on the shortcomings of conventional instruction, similar results are seen in chemistry and biology.[2] Most students are learning that "science" is a set of facts and procedures that are unrelated to the workings of the world and are simply to be memorized without understanding, and they learn to "solve" science problems by memorizing recipes that are of little use other than passing classroom exams. Furthermore, they are leaving their courses seeing science as less interesting and relevant than they did when they started.[3] The typical student is *not* learning to see science the way an expert does, as a set of interconnected, experimentally determined concepts that describe the world. They are also not learning useful concept-based problem-solving methods that can be applied in novel contexts, as experts do. Below I discuss the reasons for this and how this situation can be changed.

The Model for Higher Education: Origins and Needed Change

The current model of higher education grew in a haphazard, unplanned fashion that has left it with traditional practices and modes of organization that, in some aspects, are poorly matched to modern educational needs.[4] The lecture format, which still predominates in STEM teaching today, began before the invention of the printing press, as an efficient way to pass along basic words and information in the absence of written texts. Economies of scale led to this antiquated model expanding to the current situation of a lecturer addressing a group of largely passive students, often several hundred at a time.

Although it is doubtful that this ever was a very effective model for science education, societal changes over the past several decades have shown that it is clearly unsuitable for science education needs today. The most significant of these changes are discussed below.

Changing needs. Modern-day educational needs and goals are far different from what they were in past centuries or even a few decades ago. The modern economy demands and rewards complex problem-solving and communication skills, especially in technical fields. These skills are far more important than simple information/knowledge. The employment landscape is also changing rapidly; many current popular jobs are ones that did not exist ten years ago. The new importance of learning complex problem-solving skills is frequently at odds with traditional university teaching practices. The lecture model, while conducive to transfer of simple information, lacks the individualized challenging exercises and feedback that are critical for acquiring deep understanding and complex problem-solving skills.

Changing student demographics. Until a few decades ago, college education was necessary and useful only for a very select elite. Now college has become a basic educational requirement for most occupations in the modern economy, particularly occupations of most importance for general economic growth and personal economic success. This means that a far larger and more diverse fraction of the population is seeking post-secondary education than in the past, and thus we need a system that can deliver a high-quality education to that large, diverse population. We face an unprecedented educational challenge: the need to effectively teach complex technical knowledge and skills to a large proportion of the total population.

Changing landscape of higher education. Faculty members' responsibilities are far different from what they were several decades ago. This is particularly true at the large research universities that stand at the top of the higher-education pyramid and train nearly all higher-education faculty. The modern research university now plays a major role in knowledge acquisition and application in science and engineering. Running a research program has become a necessary part of nearly every science and engineering faculty member's activities, and it is the most well-recognized and rewarded part. Such a research program requires the successful faculty member to spend time writing proposals and obtaining research funding, managing graduate students and staff, writing scholarly articles, participating in scholarly societies, and traveling to conferences and lectures. This is much like the demands of running a small (or sometimes not so small) business. Faculty members are also increasingly encouraged by their institutions and governments to take the additional step of converting the knowledge of their research labs into commercial products. This brings additional revenue into the institutions and provides highly visible justification for government expenditures on basic research at universities. When they take this step into commercialization, faculty members are often literally running a business, in addition to having the business-management-like responsibilities of operating a university research program. While good arguments can be made for the value of these various faculty activities, the result is a faculty with new sets of demands and responsibilities that largely did not exist in the middle of the last century. These demands, and hence the need to use faculty time most efficiently, must be considered in any discussion of the future of higher education.

Growing expertise about how people learn science. While the changes discussed above affect the educational role and environment of the university, there have also been large but less conspicuous changes in our knowledge of how to assess and achieve effective science education. The understanding of how people think and learn, particularly how they learn science, has dramatically improved over the past few decades.[5] While throughout history there has never been a shortage of strongly held opinions about what "better" educational approaches entail, now there is a solid and growing body of good research supported by extensive data, particularly at the college level in science and engineering, as to which pedagogical approaches work and which do not work. These research-based methods have shown consistent benefits over the traditional lecture in many hundreds of studies across the STEM disciplines.[6] There are also empirically established principles about learning emerging from research in educational psychology, cognitive science, and education that provide good theoretical guidance for designing and evaluating educational methods and outcomes. An important part of this research is the better delineation of what constitutes expert competence in a technical subject and how this can be more effectively measured.

To briefly summarize a large field: Research has established that people do not develop true understanding of a complex subject such as science by listening passively to explanations. True understanding comes only when students actively construct their own understanding via a process of mentally building on their prior thinking and knowledge through "effortful study."[7] This construction of learning is dependent on the epistemologies and beliefs they bring to the subject, and these are readily affected (positively or negatively) by instructional practices.[8] Furthermore, we know that expert competence is made up of several features. In addition to factual knowledge, experts have distinctive mental organizational structures and problem-solving skills that facilitate the effective retrieval and useful application of that factual knowledge. Experts also have important metacognitive abilities: they can evaluate and correct their own understanding and thinking processes. Developing these expert competencies, which go beyond the factual, is part of students' path to expertness.

There are important implications of this research for both teaching and assessment. First, the most effective teaching has the student fully mentally engaged with suitably challenging, authentic intellectual tasks that embody all the relevant aspects of thinking to be learned; provides multiple ways of probing their thinking; and offers targeted and timely feedback that guides improvement in their thinking.

Second, meaningful assessment of science learning requires carefully constructed tests that measure the degree to which students have learned to make relevant decisions and solve problems like experts in a given discipline. Test design must be based on an understanding of these expert characteristics and how people learn, in addition to a thorough understanding of student thinking about the subject in question. Such assessments go well beyond the simple testing of memorized facts and problem-solving recipes that is the (unintended and unrecognized) function of the typical college examination.

Changes in the state of education-related technology. The enormous increases in the capabilities of and access to information technology provide obvious opportunities for dramatically changing how teaching is done in colleges and universities and, in the process, making higher education far more effective and more efficient. Unfortunately, these vast opportunities remain largely untapped. While there are a few spectacular examples, generally the educational information technology currently available is quite limited in both quantity and quality, in part because its design and use are not adequately guided by good pedagogy.

We are now at a watershed in higher education. We are faced with the need for great change, and we have as yet unrealized opportunities for achieving great change. Below I describe the changes and benefits that could be achieved if these opportunities were fully realized.

An Optimized University

While one might envision an ideal university that has been totally redesigned and has great resources, it is unrealistic to think that such an institution can be created. So instead I will offer a more realizable vision of a much improved university, an *optimized* university. This optimized university will provide the best undergraduate education possible within two basic constraints. The first constraint is that resources in support of higher education will not dramatically increase. The second constraint is that the long-standing structures of disciplines and departments will remain largely intact, as will current broader faculty responsibilities.

The first constraint is simply pragmatic. There is no indication that higher levels of resources are forthcoming for public education. The second has both practical and logical justifications. Where attempts have been made to create universities with dramatically different organizational structures, such as new University of California campuses without discipline-based departments, over time they have effectively reverted to largely traditional structures. I believe there is a basic organizational reason for this. There must necessarily be some organizational unit (that is, a department or some other entity) that oversees the curriculum. This unit must be able to direct the (graduate or undergraduate) career of a student based on its faculty's collective expertise as to what experiences are necessary to support student learning of the content and skills of the field. Thus, while I assume that the labels and orientation of departments will change (as fields continue to evolve because of new directions in science and technology), departments—or some similarly sized organizational entities responsible for education—will and must continue to exist. The need for entities like departments is determined by the limitations of the human brain, as there is a limit to the range of expertise that a diligent person can master. In a typical discipline or department there is a common set of knowledge and expertise that defines it. These elements are continually evolving as new knowledge and corresponding new types of expertise are found to be important for solving certain types of problems. New fields are developed and, necessarily, other aspects of expertise are dropped from the accepted canon, as they come to be seen as less important to the needs of the emerging field. For example, engineering used to be part of the physical sciences, but as engineering techniques and methods became more sophisticated, it was more productive for people focused on engineering-type problems to have a deeper grasp of those methods, at the sacrifice of areas of physics expertise. A group of people with this new set of skills thereby defined a new field of scholarship and subsequently defined what it meant to be properly educated to function well in this field. Of course, engineering itself has since subdivided into more specific fields, as the same basic process has repeated itself. In recent years the range of skills, tools, and knowledge in biology has enormously expanded, with biology departments going through a necessary process of subdividing as more specialization is needed—it is now impossible for an individual to be an expert in all areas of biology, and, correspondingly, no one individual is able to define what students should learn in order to master all areas. Thus some organizational structure like the department, which represents a defined area of expertise that one person can reasonably grasp, will necessarily continue to be the basic educational unit within the university, although the labels attached to these entities will continue to evolve with time.

Table 1.1 outlines characteristics of this optimized university, contrasted with the typical current university.[9]

Table 1.1. Differences between current and optimized universities

Current university	Optimized university
Educational focus is on the topics covered and the educational process (for example, number of students taking courses, list of topics, and so forth). Meaningful learning goals are not articulated.	Focus is on the desired student educational outcomes. Learning goals, defining what students will learn to be able to do from courses and programs, are explicitly stated.
Instructional model is that the faculty deliver information to the students, who learn it from listening and then practicing in isolation.	Instructional model is based on research on learning. Students must actively practice and develop their capabilities to become more expert, often collaboratively, with ongoing guidance of faculty members.
Faculty have sophisticated and extensive content knowledge in their discipline.	Faculty have sophisticated and extensive content knowledge in their discipline.
It is assumed that, since the faculty know the content, they know how to teach it effectively. Most faculty are unaware of the relevant research on learning and discipline-based education research.	Faculty have sophisticated pedagogical content knowledge (knowing how the content and skills are best learned, what common student difficulties are and how to overcome those, and how best to motivate students to learn eagerly and effectively).
Outcomes are assessed using tests hastily created by individual faculty members and primarily designed to rank students.	Outcomes are assessed by meaningful measures of learning collectively developed by departments, as well as student completion rates per course and program.

Academic program design	
Each academic program has a series of courses that are required. These requirements are set rather idiosyncratically and revised intermittently when someone in the department feels the inclination.	Each academic program has a clearly delineated set of educational goals that encompass the full set of skills, knowledge, and ways of thinking that are part of an education. These goals are created collectively by the faculty in consultation with other stakeholders, such as industry, educational systems, and government.
The department offers a set of courses defined by a list of topics. These choices largely reflect faculty teaching interests and past history.	Each academic program has a series of courses that are carefully aligned and sequenced to progress toward the program goals. Each course is defined by explicit and detailed learning goals that identify the full set of student knowledge and competencies provided by the course.
Faculty work in isolation to set their own agendas and goals.	These learning goals explicitly relate to the program goals and are established by a consensus of the department faculty members. They are maintained, regularly reviewed, and updated as part of the normal functioning of academic departments.
When students enter into a program, or even just a course, their background preparation is largely unknown and undetermined. Faculty members routinely spend time unnecessarily teaching known material, while also leaving large gaps in coverage of important items. Prerequisites are often poorly matched to what is needed.	The backgrounds of the students in each course will be known and conveyed to the instructor.

Table 1.1. *(continued)*

Current university	Optimized university
Teaching, feedback, and assessment	
Faculty use whatever teaching methods they prefer, usually traditional lecture.	In each class, the students encounter pedagogical approaches, materials, and technology based on careful research and measurement of results. Research-based interactive engagement teaching techniques are used extensively, with ongoing feedback provided to students (individually and collectively) during class.
Faculty members occasionally ask students which topics they have learned. A few develop review materials for the students based on their best guesses as to what some students lack that is important.	Before a course starts, students complete a detailed diagnostic examination that accurately measures their preparation. This examines their content and conceptual knowledge of the subject and those subjects that the course builds upon, such as mathematics and related science disciplines. It also diagnoses their attitudes and epistemologies about the subject and how it is best learned.
Many faculty members spend several class periods rapidly reviewing the knowledge they think the students need. The students who already know the material are bored and find this a waste of time, and often they become overconfident as to the level of challenge of the course. The students who are less prepared don't benefit because the review is too rapid to learn from.	Before a student has ever seen an instructor, the instructor has a profile of his or her strengths and weaknesses, and the computer has already flagged serious deficiencies. If these deficiencies are widespread, the student is guided to enroll in a more appropriate course.

	Where the deficiencies are localized and not severe, the computer provides the student with feedback and suitable exercises to complete that remedy these deficiencies. This ensures that the course will begin with all students at roughly the same level of knowledge and competence, and the instructor has an accurate profile of that level.
	The instructor uses the profile of the class to suitably tailor the learning environment.
The frequency of evaluation of the students is determined by the instructor and typically includes only graded homework (although often does not), one or two midterm exams, and a final exam. Due to a lack of faculty expertise, many homework problems and faculty-created exams primarily practice and test memorized facts and procedures. Feedback from these evaluations is usually delayed by one to two weeks and provides little to the students beyond a score showing the fraction of questions answered incorrectly.	There are regular ongoing evaluations of the student's thinking and learning throughout the course. These evaluations are linked to targeted timely feedback to both student and instructor. Information technology is used widely to support this ongoing evaluation and feedback, including online homework systems that include intelligent grading and tutoring programs.
Typically, technology is developed and used for its own sake, often provides little educational value, and is seldom evaluated as to its effectiveness.	Technology is chosen by looking carefully at how it can enhance learning by supporting good pedagogical design, enhance the capabilities of the instructor, and improve instructional efficiency.
Collaborative learning by students is discouraged by curve grading and relies on informal student arrangements. Communication and teamwork skills are usually not part of regular science courses.	The full benefits of collaborative learning are realized by building such collaboration into the structure of the classes, assignments, and grading. This also improves students' teamwork and communication skills.

Table 1.1. *(continued)*

Current university	Optimized university
Faculty teaching evaluations are based on student course evaluations. These have little correlation with learning and none with the use of effective teaching methods.	Faculty teaching evaluations are linked to good measures of student learning and the use of the most effective teaching practices.
Saving faculty and student time—improvement in efficiencies	
Each new instructor to a course typically reinvents it anew, spending a huge amount of time in the process, and creating a course that is informed neither by past experience at the instructor's own institution nor by experience at other institutions nor by relevant educational research.	Benefiting from experience and saving time by copying what works, course materials and teaching methods are passed from one instructor to the next and continually improved. The relevant discipline-based education research is consulted and good examples of teaching the topics at other institutions are copied.
Teaching is an isolated activity in which faculty set their own agendas and goals for the courses they teach. Topic emphasis, teaching methods, and assessments vary widely depending on who is teaching. As a result, there is limited coherence within the curriculum and considerable inefficiency in coverage, wasting both student and faculty time.	Collective ownership is felt within department for courses and curriculum. Collective ownership, along with use of best practices listed above, ensures consistency, coherence, and effectiveness across the curriculum, saving faculty time in preparation and instruction, and making student learning greater and more efficient.

As noted above, the range of student preparation in every course, and the lack of information as to that preparation, results in teaching that wastes large amounts of faculty and student time due to instruction that is redundant for many while being too advanced for others.	Spread and uncertainty in student preparation is reduced. As discussed above, use of diagnostic exams and targeted interventions, and greater educational effectiveness of courses and consistency in learning guided by clear learning goals, will greatly reduce spread in student preparation and provide detailed information to the instructor for every class.
Because of uncertainties as to coverage and learning in previous courses and general student preparation, the same science topics are covered repeatedly in the curriculum of a science major, but commonly covered so rapidly each time that the students do not achieve mastery. Such a process is inefficient as well as pedagogically flawed. The research shows that if students develop an incorrect understanding of the material, then shallow repetition tends to reinforce rather than correct such misunderstandings.	Unnecessary repetition is avoided. As discussed above, because of the carefully designed coherence in the curriculum, coverage in each course is optimized to build on what comes before it in the most efficient manner.
A large amount of faculty "teaching" time is spent on low-level administrative tasks that could be performed by less expert and lower-cost staff. This involves routine class maintenance, recording of grades, dealing with students who are dropping or adding classes, dealing with special student circumstances such as missing assignments or exams due to medical or family emergencies, and so forth. For large classes these low-level tasks can take up a large amount of faculty time.	Expensive faculty time is no longer spent on low-level administrative tasks, particularly in large courses. Dedicated administrative staff, aided by effective software, handle all such tasks more effectively and at lower cost than when done by faculty members. This includes making arrangements for the increasing number of students who have special needs that require adjustments in teaching and/or testing.

Table 1.1. *(continued)*

Current university	Optimized university
Just as faculty receive inadequate training for teaching, so do TAs. TAs are often put in charge of lab and recitation sections and given very little guidance or oversight. A few exceptionally dedicated TAs provide a good educational experience for the students, but the majority do not develop sufficient expertise.	Trained teaching assistants become important contributors to undergraduate education. Well-designed and tested training programs routinely produce extremely well-qualified TAs. Both graduate student and undergraduate student TAs develop a good understanding of how to effectively address student difficulties within the discipline. This provides excellent educational experiences for students and also is the first step in developing science teaching expertise for future faculty and K-12 teachers. This is a low-cost way to improve the student educational experience.
Class size is determined by one of three factors: (1) the economics of maximizing the number of students per instructor, (2) a general unsupported belief that there is educational benefit to smaller classes, or (3) the faculty preference for teaching smaller classes, particularly in their areas of specialization, because such classes are more fun to teach, with more personalized interactions with students, and there is less administrative burden with fewer students. This often results in departments creating many small boutique upper-division courses, along with having very large introductory courses. There is no reason to believe that the current class sizes are optimal in any educational or economic sense. Costs per credit hour are usually heavily weighted toward upper-division courses instead of lower-division courses, when good arguments can be made that the opposite would be	Class size is set based on careful measurement and optimization of educational benefits per cost. Technology is used to enhance the most effective research-based interactive teaching in all classes, but it provides the most benefit in larger classes. Technology continues to be developed to help the instructor to make larger classes more intellectually engaging, personalized, and educationally effective. Research is not available to say what class size optimizes learning given a fixed amount of resources and best instructional methods, but the mantra of "smaller is better" is almost certainly not the optimum. There are demonstrations of classes of 200 or more achieving very good learning gains by utilizing technology and research-based practices—with skilled instructors, nearly as good as the best that has been demonstrated in much smaller

Issues and Challenges in Optimizing the University

There are some substantial impediments to moving from the current situation to the optimized university. These include structural and administrative limitations, the balance of research and teaching, and failures in the market and in incentive systems.

Structural and Administrative Limitations

University governing systems are poorly suited to making changes on a time scale that is rapid relative to the faculty life span, which can be several decades. The tendency in the United States toward rather short-lived upper administration (the tenure of public university presidents in the United States now averages less than five years), combined with the pattern of sharing governance with faculty members who have careers lasting decades, effectively puts the administration in a very weak leadership position. In the United States, university governing boards and the position of public university presidents have become highly political and subject to the vagaries of current events, college athletic teams' success, and political intrigue, thereby greatly weakening and distracting administrative academic leadership. Unfortunately, at the same time that administrative leadership is being weakened, modern research universities have grown too much in size and complexity for regular faculty to have all the information and experience needed to make major institutional policy decisions. Faculty members simply do not have the time to become sufficiently aware of all the issues and pressures, but they remain a powerful entrenched body that can hinder change. This combination of factors reduces the organizational capacity to carry out useful long-term strategic planning, investment, and implementation of desired changes, such as the optimization of undergraduate education described above.

Another closely related complication is that that actual "ownership" of educational activities rests almost solely within departments. Realistically, this is necessary. It is impossible, for example, for someone with a background in history, or even in a science such as physics, to be able to say what students should be learning in their biology classes. However, this also means that educational change must happen at the departmental level—it is very difficult to mandate it from a higher level and achieve the desired effect. Thus educational reform efforts almost certainly have to be based on a model for change at the departmental level.

Balance of Research and Teaching

The appropriate balance of teaching and research in the optimized university remains a matter of debate, with no clear best weighting. Both teaching and research are essential components of the modern research university and are vital contributions to society, and to be a highly effective teacher in a discipline, one must be an expert in that field (as well as having expertise in teaching). It would be unwise to abandon either. However, optimizing the use of faculty time offers enormous potential for improvement in educational effectiveness and efficiency. The best approach is to achieve those improvements and examine the results before considering any changes to the current balance of research and teaching. Also, it is hard to imagine that faculty members could teach expert competence in an area of modern science and technology unless they have been active in the field themselves for much of their careers. The complexity and rapidity of progress in these fields today are such that faculty simply cannot remain sufficiently expert in the subjects in which they are educating students if they must rely on teaching the subject based only on what they themselves learned in school. Thus maintaining an active research program in a department clearly serves to enhance the desired faculty expertise in teaching.

Failures in the Market and in Incentive Systems

Teaching in the modern university displays a well-known phenomenon in economics: that free markets do not function properly in the absence of information. In the context of higher education, it is next to impossible for prospective students to get any meaningful information on the quality of teaching at the institutions they are considering. So they are forced to make decisions based on very distant proxies, such as the research productivity of the faculty at a given institution, the cost of tuition, or the quality of the dormitories. Once at the institution, they might be able to make decisions about courses based on student course evaluations, but it has been well established that such evaluations have a host of problems, the most important being that there is no correlation between student evaluations and objective measures of learning,[10] and we have seen no correlation between evaluations and the use of effective instructional practices.[11]

As a result of these information failures, the educational value provided by an institution of higher education, how sought-after it is by prospective students, the amount of public support it receives, and support provided to the faculty who generate that educational value are all completely disconnected. The lack of information results in a lack of incentives to improve educational quality.

The biggest barrier to improving the teaching at research universities is that they are so ineffective at measuring and rewarding effective teaching. There are no incentives for educational change built into the system, and there are several disincentives. Only after the lack of effective measures of teaching quality is addressed will it be possible for prospective students, state governments, the public, and institutions themselves to recognize and reward teaching quality. This will provide the necessary incentives for institutions, and faculty within those institutions, to adopt the best teaching practices and work to improve educational outcomes. We have developed better methods of evaluating teaching as part of the SEI efforts.[12] When measures such as these are in widespread use and the resulting information is available, it will then be possible to have a meaningful incentive system that will drive ongoing improvement in educational quality. This will also allow rational decisions about the appropriate weighting of research and teaching in the optimized university, as well as sensible variations in this weighting across different types of institutions.

TWO

The SEI Model for Achieving Change

THE GOAL OF the SEI was to improve undergraduate science teaching, but this required change in established traditions, practices, and cultures of research-intensive universities, which are inherently large, complex organizations. I recognized that this was a formidable task and put considerable thought into the design of the initiative, attempting to craft a model that would address all of the most critical factors. This required first identifying as many of the important factors as I could, which I did by talking to many people and looking into the research literature on both adoption of innovations and bringing about change in large organizations. Early on, it was clear that the academic department was the critical unit for changing teaching at such institutions, as departments control what and how the science courses are taught. So, I did my best to identify the values, beliefs, and practices (that is, the "culture") of each different science department and to see how the general principles for achieving innovation and change would apply in that context. These considerations led to the model for change represented by the SEI and discussed in this chapter. It was intended to apply leverage for change at the most essential points and address all the critical barriers to adoption of novel teaching methods, while recognizing that there were many unknowns.

In the current culture of university STEM education, the impetus for improvement relies primarily on individuals acting alone, rather than on organizational structures supporting that change. As a result, teaching innovations are inherently fragile and challenging to scale up throughout an institution. The focus on individual creation of instructional materials is also inherently inefficient, as faculty continually reinvent the wheel. Currently hundreds of instructors each year individually invent their courses anew, even though the equivalent course is and has been previously taught in many other institutions, including their own. However, the knowledge and materials produced by all those other examples, as well as the research on more effective ways to teach specific topics, are not being widely shared or used. A new institutional culture is needed that supports coherent, collective efforts to use the most effective teaching methods and optimized instructional materials. Organizational structures and incentives also need to support this culture.

The goal of the SEI was to transform undergraduate science education by creating a culture within academic science departments where research-based, effective teaching and course design were the new normal. The SEI focused on the department as the essential unit for educational change and on the large public research university as the most relevant institutional type. The core component of the SEI model was that departments competed for substantial one-time funding to support changes in teaching, with most funds being used to hire postdoctoral education specialists to work with faculty within the department, and the remainder going toward direct incentives to faculty. This structure provided expertise, skilled labor, and incentives for educational innovation, offered support for a limited time in order to create a sense of urgency, and helped forge shared visions for change through the development of proposals for the competition. The desired outcomes were improvements in course design and student learning, improved faculty teaching expertise, shared course resources, and an overall cultural shift in departmental norms for instruction.

I started with a plan for how courses might be designed to be more effective, based in part on my own experience in successfully transforming some physics courses. These transformations started with articulating a detailed set of learning goals, then creating instructional activities for class and homework that targeted these goals and were based on methods that research had shown to be most effective. I created multiple ways to measure how well students were achieving these goals, and used these measurements to optimize courses through multiple offerings. In these efforts, I was assisted by Katherine Perkins, a talented recent PhD in chemical physics who was interested in developing expertise in teaching. I saw how enormously valuable it was to have a collaborator such as Kathy helping with these course transformations. These courses were subsequently passed on to other instructors who continued to teach them using many of the same methods and materials. I was also inspired by work at the University of Illinois in which a departmentally owned large introductory physics course was established, with faculty members rotating in as part of a team to teach the course, using established materials and research-based teaching methods.

Achieving widespread change in educational practice, as described above, involves changing both the individuals involved in teaching and learning and the academic organization that represents the aggregate of these individuals, along with the procedures, cultures, and norms of that organization. This brings together aspects of both diffusion and adoption of innovations. In this case the innovation is more-effective teaching methods. In universities, the academic department is the dominant organizational unit with regard to education, with larger institutional structures exerting an important but distant and rather diffuse influence. The SEI model was guided by the literature on what factors facilitate and inhibit the spread of innovations and organizational change, particularly the work of Everett Rogers on the adoption of innovations and the work of John Kotter on organizational change. The principles presented in those works were, to the extent possible, implemented in the context of large research-intensive science departments at large research universities.

The Diffusion of Innovations in Education

Rogers has laid out five steps (see Figure 2.1) that individuals and organizations go through sequentially in the successful adoption of innovations: knowledge, persuasion, decision, implementation, and confirmation.[1] At each stage there can be failure and, consequently, uncertainty as to whether the next stage can be reached. These stages offer useful ways of thinking about how to bring about innovation in undergraduate education.

First, one must have some mechanism to increase knowledge: the level of faculty awareness of alternative types of pedagogy and of research on learning. Next is persuasion: convincing them to learn more about the innovation. The third stage, decision, involves establishing an environment in which faculty perceive a benefit-to-cost ratio that is sufficiently favorable that

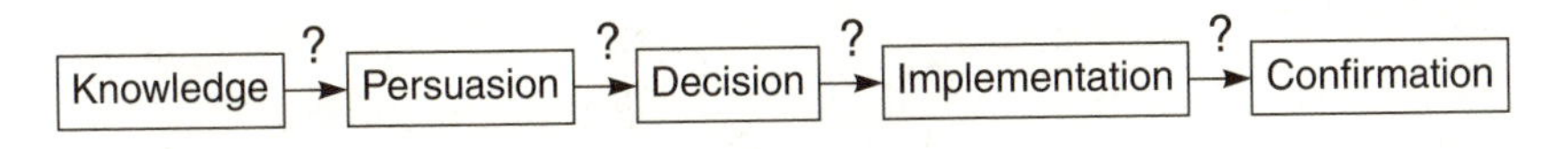

FIGURE 2.1. Steps in the adoption of innovations
Source: Everett Rogers, *Diffusion of Innovations*, 5th ed. (New York: Free Press, 2003).

they will decide to adopt, or at least tentatively try, the innovation (new ways to teach). That is followed by implementation, a critical stage in which they try teaching differently and decide if it is successful or not. That finally leads to the confirmation stage, in which their experiences with these new approaches lead them to decide whether they will continue to use the innovation. In this case, that experience includes their personal feelings and the feedback they receive from students and the department.

In many work practice innovations, it has been shown that the success of a change process often depends heavily on how it is related to culturally based practices of the organization and how it impacts core members' self-identities.[2] As Rogers discussed, what seems to matter most in individuals' attitudes and responses to proposed innovation is the way in which they perceive the relative value of any change—that is, whether and how they can link what is proposed to what they already value.[3] There are two rather distinct aspects of the culture of a science department at a research university: the culture of teaching and the culture of scientific research. The goal of the SEI was to change the teaching culture, but to carry out that change in a way that relies heavily on the values and practices of the research culture. This shift, I hoped, would bring the teaching and research aspects of the culture much closer together, which should facilitate the change process.

Faculty members who do scientific research understand and value quantitative results. Also, faculty understand and value conceptual and higher-order thinking skills and expert attitudes about science. Thus, the SEI aimed to provide faculty members with meaningful ways to assess student learning, particularly higher-order thinking skills; to show that these assessments quantitatively demonstrate the superiority of new research-based teaching methods over traditional approaches in terms of getting students to think more like scientists; and to show that there are underlying empirically determined principles of learning that can be used to design instructional activities and provide predictable results.

Essentially, this model would have the self-identity of faculty members as scientists expand to include their identities as teachers of science. However, this requires that their teaching practices and measures of success be based on research, empirically grounded principles, and objective data. Although this was the original design concept for the SEI, I learned that it gave too much emphasis to faculty as scientists and the belief that their "scientific thinking" would transfer over to how they thought about teaching. In reality, while there was a complex mixture of reactions, teaching was generally viewed more as a personal, emotion-based activity than as a scientific, evidence-based activity. During the vital persuasion and decision stages, it turned out that the dominant factors for most faculty were the personal satisfaction and emotional responses they received from teaching and from interacting with students in a particular manner. This was balanced against the feedback they felt was provided by the formal incentive system in terms of their research productivity and how they were evaluated by students. That formal incentive system was entirely negative to innovative teaching, but what mattered was how negative it was perceived to be.

There are many other factors that can influence faculty and departments in their decisions to try or reject educational innovations. As discussed in the next section, the SEI attempts to address most of these, starting with providing large amounts of flexible money to departments.

Once faculty members and their departments are committed to trying to transform and improve the undergraduate education they provide, there are still three significant hurdles that are evident when one maps Rogers's stages model onto a faculty member's adoption of innovative teaching methods. First, typical science faculty members have little knowledge of research on learning, of meaningful assessment techniques, and of effective research-based teaching practices. Second, they do not have time to go out and learn these things on their own, let alone put them into practice effectively in actual courses while maintaining their current level of other responsibilities for research and service. Third, most do not have knowledgeable, interested colleagues with whom they can discuss and develop these novel teaching ideas.

Unlike science research, science teaching is typically a solitary effort. Many teaching improvement efforts have involved the formation of "learning communities" devoted to development and implementation of improved, innovative practices, and I too wanted to establish teaching as much more of a collaborative process among faculty. Such collaboration is also an essential part of the scientific research enterprise, and so by building this into the SEI model, I again aimed to incorporate cultural values from faculty members' scientific research identities.

A second classic aspect of Rogers's work is the classification of the members of an organization considering an innovation into five groups: innovators, early adopters, early majority, late majority, and laggards. While some aspects of this classification scheme are convenient, I found it was not very useful for characterizing the adoption of innovative pedagogy by faculty, because, as discussed in "Faculty Attitudes about Teaching" (Chapter 5), individuals often do not fit well into such a simple categorization of attitudes, particularly when examined over multiyear time scales. There certainly are a few who are much more willing than others to try out new teaching methods, and a few who are quite resistant, but beyond that, things get more complicated. Some individuals are early adopters of a particular aspect of pedagogy but then are quite resistant to more extensive changes, while others may come to embrace novel pedagogies slowly but do so in a much more deliberate and extensive way. Also, predictions about the later behavior of individual faculty members based on their early reaction to innovative pedagogy (or their age or other factors) often turned out to be wrong.

Lessons from Recent History

In considering the goals and model for the SEI, it is useful to examine one recent example of a large and rapid change within universities: the enormous growth in the university research enterprise after World War II. As a result of this change, research is now an essential component of every large university and provides a major service to society. Most public U.S. universities shifted from being predominantly institutions focused on teaching students of their respective states to being modern research universities that looked to the nation and the world as their stakeholders.

There were three key factors in this change: (1) the shift was largely faculty driven, (2) there were clear measures of success, and (3) there were clear incentives for change at both the level of the individual faculty member and the department level. Individual faculty members saw that external research funding had become available, and they recognized that this would allow them to do more science, which in turn would increase their status both locally and among the wider community of scientists in their discipline and allow them to contribute to society in new and important ways. Transformation happened at the department level because departments primarily determine faculty hiring, review, and salaries, and the values of the department fuel or inhibit change in how faculty spend their time. There were clear incentives to departments to encourage faculty research activities (increased funding, larger and better facilities, increased prestige, better students), and there were clear measures of outcomes (research dollars brought in, papers published, work cited, scientific awards, departmental rankings) that became collectively accepted. These outcome measures became embedded in departmental and institutional evaluations, reward systems, and hiring criteria. This in turn drove the job market to give higher priority to potential faculty members who were more successful according to these measures. The resulting market forces impacted all colleges and universities. To hire good faculty, it was necessary for an institution to encourage and support research activities. The outcome was a major transformation of universities, largely driven by entrepreneurial faculty who saw clear incentives for their efforts in the large amounts of federal research dollars that had become available. While support and encouragement from the higher administration was important, the change was carried out at the levels of departments and individual faculty members.

This example suggests elements that are important for pursuing any widespread change in the university context, in this case the change being how the science courses are taught. Incentives to individual faculty members and departments must be clearly tied to educational outcomes under their control. Relevant outcomes must be readily measurable and show what is needed to achieve improvement. In addition to allowing comparisons between individuals, the outcome measures must also allow comparisons between departments and between institutions. The evaluation, reward, and hiring policies of the department and the institution must reflect the desired outcomes. And faculty who are successful by such measures need to be able to achieve greater recognition within their discipline, such as through publications, conference talks, and awards.

Putting all of these elements in place to improve STEM education will be a difficult and long-term challenge. However, it is much easier for an institution to implement the smaller set of elements necessary to drive department-wide improvements in teaching and to facilitate the efficient adoption of those improvements within the institution. The SEI was an experiment in trying to attain that goal.

Components of the SEI Model and Guiding Principles

The SEI was composed of the core components shown in Figure 2.2.

The adoption of innovation discussed above needs to take place in the context of an organization—individual academic departments and, to a lesser extent, the broader university. Few of the relevant decisions are being made by individuals in isolation; rather, they are shaped by the organization in which those individuals work. One must consider what is necessary to change the organization so that it encourages and supports the use of innovative teaching. The initial design of the SEI was based on many of the elements Kotter identified as necessary for organizational change to succeed, as applied to the context of research-intensive university science departments.[4] It should be said here that while experience supported the validity of all the important factors listed by Kotter, I was unable to successfully address all of them in this context; this is discussed further in Chapters 5 and 6.

My first guiding principle was that the SEI was to be a one-time, limited-duration infusion of resources to change practices and culture that would

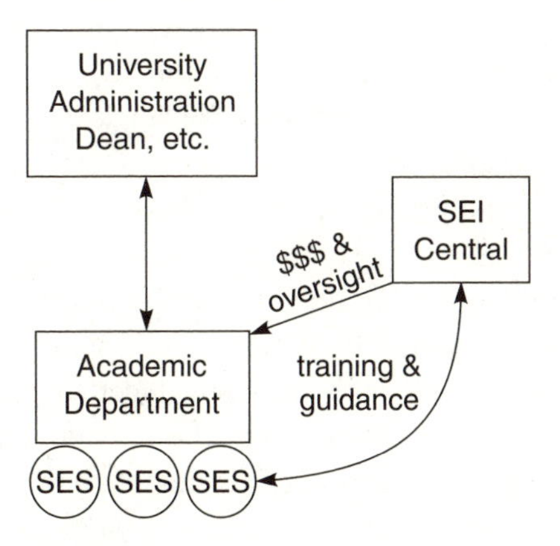

FIGURE 2.2. Core components of the SEI

A competitive grant program invited departments, not individual faculty members, to compete for substantial one-time funds. Several departments were funded. SEI Central made the decisions on funding and provided oversight to the departments that received grants. It also provided training and guidance to the science education specialists. Science education specialists were hired by the department with SEI funds. These provided expertise in teaching in the discipline, and also worked with faculty members to transform courses and teaching and to assess the results according to the SEI model.

then become self-sustaining. It costs money to bring about change, but the expectation was that the long-term ongoing costs of instruction would be the same as or less than what they had been prior to the SEI. The spirit was much the same as investing in the cost of retooling a factory with better equipment so that it can produce a better product at the same cost as before.

The scale of funding needs to be commensurate with the scale of change expected. The organizational change literature (largely based on studies of industry) indicates that major changes involve investments of 5 to 10 percent of the annual budget of the organization for time scales of around five years. I estimated that 5 percent of the annual budget of a large science department was about $400,000, and if that level of support was provided for five years it would come to a total of $2 million. A period shorter than five years would not be realistic for the scale of change that I was attempting, but a longer period would make it easy to put things off. This amount would cover the estimated costs of the labor involved in transforming the twenty-five to thirty undergraduate courses regularly offered by a large department. This meant that about $10 million was required for five large departments, to be spent over a period of about five years. This was a factor of ten to a hundred times larger than typical federal or institutional grants provided to improve teaching in the early 2000s, when plans for the SEI were being formulated, as grants typically targeted single courses or single individuals.

If there was to be any hope of making change sustainable, it had to involve a substantial number of the science departments at an institution. I chose five (of about eight) as the optimal number. If that many departments carried out major change, it would likely establish new norms for teaching science at the institution. And because the only model that science faculty and departments are familiar with for coming together and formulating consensus plans and commitments involves the pursuit of large competitive grants, I decided that funding for the SEI should be through a competitive grant program to which departments (not individuals or collections of faculty) could apply, and the chances of receiving funding needed to be high enough to warrant serious collective effort but low enough to give the sense it was a real competition that required their best effort. I also stressed the importance of experimentation and collection of data.

Because departments need to feel ownership of the effort and the changes that result, it is the departments themselves that must initiate participation, deciding as a unit whether to submit a proposal. This structure is designed to create a scenario in which departmental faculty collectively discuss SEI participation and the majority have expressed a desire and commitment to engage in improving science learning.

There needs to be a meaningful incentive for people to put in the effort and time required. This is true both for individual faculty members with regard to changing their teaching and for the department administrators with regard to the oversight of these changes.

The transformation of courses and the development of a sense of collective ownership of courses will occur only if the faculty's teaching methods and level of knowledge about teaching are transformed as well, so the processes of course change and change in individual faculty members' teaching should be integrated. There should be a specific structure to the course transformation process and specific outcomes for a transformed course, to ensure appropriate guidance and deliverables. As the process develops, highlighting early successes and small wins will build interest and enthusiasm.

Departments seldom have the necessary expertise in teaching and learning, but for long-term success such expertise must reside in the department. So the program needed to find a way to introduce it and embed it into the departments. Use of science education specialists (SESs), who are well grounded in the discipline and knowledgeable about teaching and learning, working with the faculty was the proposed mechanism for achieving that growth of departmental expertise. Having them be junior to the faculty has benefits, as the specialists will be more inclined to work with faculty in a partnership, rather than telling them what to do and being annoyed if their recommendations are not followed, and they are more willing than senior people to provide labor.

It is neither possible nor desirable to try to change everyone at once. The design was to systematically support the change of teaching by a fraction of the faculty each year, starting with the early adopters. The original concept of the SEI was that a department would systematically work to transform its undergraduate courses, starting with the introductory courses and then progressing up through the undergraduate program. For a variety of reasons discussed in Chapter 6, this approach did not work. As a result, I abandoned the idea of having departments change courses in a logical order and instead focused on ensuring departments had good planning and incentives in place to maximize the number of faculty fully engaged in transformation efforts, and to maximize the number of courses transformed.

The greatest barrier to faculty's changing their teaching is the time it requires. In order to make changes, faculty must use time that would normally be spent on research. As Kotter says, a sense of urgency—the feeling that this needs to get done now, and so it must take priority over the countless other demands on faculty members' and department chairs' time—is very important. As I will discuss in more detail in Chapters 3 and 5, generating such a sense of urgency always proved challenging, and over time I came to realize there were some unique features of education in the university setting that were responsible for this. As a result, I added some requirements for funding that modestly helped to encourage a sense of urgency about the SEI-supported activities.

Finally, because I was sailing in uncharted waters, I knew that considerable flexibility was needed. I had to be ready to make changes and adjustments based on what was working and what was not.

These components and principles were intended to address Rogers's first four stages in the adoption of an innovation, as well as the factors identified by Kotter as important for organizational change. I recognized that they did not address the longer-term question: assuming the changes were successfully implemented over the study period, would they become part of the culture and be sustained after the project's conclusion? I hypothesized that they would in fact be sustained, because the individual faculty would find that teaching this way was far more personally rewarding, the departments and higher administration would see compelling improvements in student learning, and the faculty and departments would value the gains in efficiency provided by collectively owned and systematically optimized and shared courses. Although more time is needed to determine if the changes produced by the SEI will be sustained, the results have been mixed so far. Only the first factor, greater personal satisfaction from teaching, has been realized, but it is proving to be more powerful than I had previously thought.

Different Institutional Contexts

The science education initiatives were separate programs with similar designs at two fairly comparable universities. Both were large public research-intensive universities that were the most prominent institutions in their respective geographical regions. The University of Colorado (CU) is the most prominent research university in the sciences in the Rocky Mountain and western Great Plains region; the University of British Columbia (UBC) is the most prominent university in western Canada. There is a great deal of similarity between these two universities at the level of individual faculty members and departments, and most UBC science faculty members have spent time in U.S. universities. I also found the general structure of the curriculum and the cultural beliefs of particular disciplines about learning and teaching to be very similar—for example, the math departments, physics departments, and chemistry departments showed far more similarities with their counterparts in the other institution than they did with other science departments within their own institution. Demographically, the UBC student population is somewhat more diverse than at CU and is majority Asian.

The original plan for the SEI was to achieve economies through the sharing of materials, data, and infrastructure between the two institutions. As noted in Chapters 5 and 6, very little of this happened. There turned out to be relatively little overlap between departments supported at the two universities, and there was also not much overlap between specific activities within similar departments. Also, as noted later, a failure of the SEI was that few of the efficiencies achievable through the sharing and exchange of materials and efforts were ever embraced by faculty at either institution. However, there was considerable sharing of experience and wisdom with regard to the best ways to structure SEI funding and management and training programs, and there was some sharing of ideas and methods between SESs at the two institutions.

There were various institutional differences that had to be considered. One was the stability of the administrations. At UBC, the administration at every level had been very stable, with administrators serving out their full five-year terms (and sometimes going on to serve a second term). At CU, for decades the turnover at all levels had been much higher and usually turbulent, making it much more difficult to imagine any large-scale institutional change driven by the administration.

A difference that turned out to have little impact was nomenclature: CU has department chairs, while UBC has heads. In all cases at both institutions, the authority and effectiveness of the head or chair seemed to be determined by the person's skills and stature in the department rather than by any formal authority. For simplicity, in the rest of this book I will just use the label "chair."

A more important difference was administration involvement with the SEI. At UBC, the SEI was a highly publicized activity, with both the president and the provost participating in a number of events where it was highlighted; it was also the subject of a number of high-level university meetings, and there were regular reports on it to the Board of Trustees. The dean and relevant associate dean were involved on a regular basis, typically meeting monthly to discuss progress and on multiple occasions intervening with department heads when problems arose. The dean often spoke about it in public events as a point of institutional pride. Raising money for the extension of SEI-type activities after the original funding ran out was made a priority by the dean, and the dean was a prominent presence at the annual SEI mini-conference. Perhaps most important, the dean ensured that when new department heads were appointed, they were supportive of the SEI.

At CU, there was no significant involvement by the administration beyond the initial funding. Annual reports on the progress of the SEI were provided to all levels of the administration each year, but there were never any responses or follow-up discussions of these reports. In the selection of new department chairs, there was little if any consideration given to their attitude toward the SEI.

Although the authority of the dean was more limited at UBC than at CU because of how the institutions handled budgeting and faculty salaries, over time the difference in the deans' support of the SEI could be seen to have substantial impact, largely through the choices of appointments of department chairs and the messages implied by those selections (discussed further in Chapter 5).

A fourth institutional difference was that there was considerably less accountability at UBC on the level of individual faculty members. CU faculty have to complete a lengthy annual performance report documenting their research, teaching, and service activities, and departments and the dean then rate the research, teaching, and service performance of each faculty member, which determines a substantial fraction of the annual salary increment. At UBC, the faculty is unionized, and salaries and annual raises are almost entirely determined through a collective bargaining agreement. Faculty only submit an optional and informal report on their performance if they want to be considered for the very small fraction of the salary increment that is based on merit. At both institutions, the evaluation of teaching at the institutional level is predominantly based on student course evaluations and was perceived to have little weight.

Yet another institutional difference was that UBC is the institution of choice for students in British Columbia, a province that by international comparisons has a very good K-12 education system, and most Canadian students do not move around the country to go to a university. As a result, the students at UBC, particularly in the sciences, are better than at CU on average, but there is a large overlap of the two distributions. Curiously, when the SEI started, there was a pervasive and frequently expressed sentiment among the faculty at UBC that the students were weak, either in their academic preparation or in their work ethic, and that many of them did not deserve to be at UBC. Such sentiments were expressed far less frequently at CU, and the origin of such opinions at UBC was difficult to understand. However, there are hints that as teaching methods have improved and become interactive, such faculty sentiments may also be changing.[5]

There had already been a number of improvements in the teaching of science at CU before the start of the SEI, and there was generally a greater awareness and use of research-based teaching methods there than at UBC. There also was a relatively strong effort at CU in discipline-based education research (DBER) in physics and biology, with smaller efforts in other departments. These activities had been largely spearheaded by myself and other prominent science faculty members. The impact of the DBER program on the SEI work is unclear. At one level it provided a greater knowledge base and enhanced capabilities for assessment. However, I also got the impression that it created a sense among faculty members that "improving teaching is the job of the DBER faculty, and so it is not my responsibility" and thereby diluted efforts. There were also times when it appeared that the condescending attitudes of some DBER faculty may have made some regular faculty less inclined to be involved with innovative teaching methods. Over the course of the SEI, there was substantial growth of DBER at UBC.

A sixth institutional difference was that the overall funding models for the two universities are different, and while both are complex and have somewhat different priorities and constraints, it appeared to us that UBC was somewhat better funded.

A related difference was that just before the start of the SEI there had been a series of budget cuts at CU due to reduced state support, and so providing the $10 million needed for the SEI at CU would have required conspicuous cutting for other programs. This would have hurt other aspects of education and would likely cause substantial resentment among faculty and departments. That financial reality led to negotiations that resulted in a $5 million SEI at CU with a substantial fraction of that not coming from general funds, while the UBC program had a commitment of $10 million.[6]

That led to one last difference in the two institutions: because of the amount and nature of the funding, there was significantly less SEI Central support at CU. The CU director had a 20 percent appointment, the associate director position was a half-time appointment, and there was a 50 percent administrative assistant. At UBC the SEI was staffed at about twice this level. While it was intentional to have a very lean central staff and invest as much as possible in the departments, the staffing at CU was too lean (see Chapters 5 and 6), particularly after I left the director's job to take a position in the White House, leaving the associate director to take over those duties as well.

THREE

The Process of Making Change

IMPLEMENTING THE SEI across two institutions and many different departments involved putting in place many pieces and finding ways to adjust those pieces when unanticipated challenges arose over the course of six years. This chapter describes that full process of implementation, starting with the method for funding the work and then putting in place the planning, departmental and institutional structures, and oversight required to move the project forward. I then discuss the core of the SEI effort, how the faculty members in each department were supported in a deliberate process of transforming the courses they were teaching, and in that process, transforming their approaches to and methods of teaching. The final part of the implementation was collecting as much data as possible as to the results, including the differences across institutions and departments.

Implementation of the science education initiatives spanned more than six years, with many tasks having to be completed within the first year. (See Figure 3.1.)

Proposal Process

When departments first considered the SEI call for proposals, the concept was so novel that they had little idea of what to do; thus the proposal development process was fairly interactive. During the proposal development process, departments were provided with a framework for carrying out

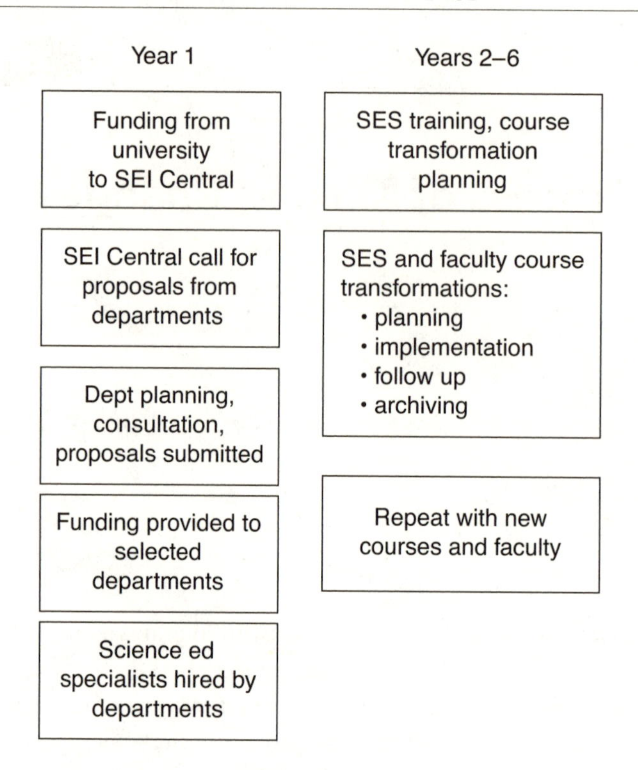

FIGURE 3.1. An implementation plan spanning six years

changes, including the vision of a transformed course and the possible use of SESs in this process. This framework guided departments in the types of activities that they could support with SEI funding. While the framework encouraged a general set of activities, the focus was on the outcomes, and departments had substantial discretion in how they expended the funds and how they structured their proposed work. The departments were explicitly told that funds could be carried over from one year to the next, to optimize how they were spent. Additionally, as a result of early experiences, it was important to make it explicit that funding could be discontinued in future years if sufficient progress was not made.[1]

In all cases, shortly after the call for proposals was sent out, I would attend a departmental faculty meeting to discuss the research on science education, what they might do to improve undergraduate education, and how to go about it. The proposal process and decision-making criteria were also presented. In retrospect, the level of faculty participation, the issues raised, and how the chair managed the discussion and dissent (primarily about the effectiveness of different teaching methods) during those early meetings turned out to be a fairly good predictor of the later outcomes of the department. During this process, either in those meetings or in materials provided to the department, the department would be introduced to the idea of SESs who could be hired with these funds and trained by SEI Central. Because the departments had so little precedent for an effort like this, they had difficulty understanding all that would be involved, and hence had difficulty being very specific in their proposals.

Early experiences led to adding the requirement of an explicit list of courses to be changed, a roster of faculty who would be involved, and a timeline for the project.[2] When these requirements were not put in place, many commitments were largely ignored after funding was provided. Making the commitments more specific helped to ensure that the department carried out adequate planning and was ready to live up to the commitments it was making, and helped to add some sense of urgency through having milestones and timelines. Even if these timelines were not strictly adhered to, having such goals resulted in fewer problems within departments.

One very early success was that the SEI call for departmental proposals resulted in all of the departments at both institutions having serious department-wide discussions about how they might improve teaching in their undergraduate courses. Although departments have often had discussions about curriculum, in most (and quite possibly all) cases this was the first time department-wide discussions about pedagogy had ever taken place. When I spoke at the faculty meetings about research on how to improve science teaching, this was the first introduction of these ideas to most of the faculty. To encourage such department-wide discussions, one criterion was that the departmental proposal had to be submitted to a faculty vote.

Resistance to the SEI efforts also emerged early. One expected source was a set of faculty members known to attach little importance to undergraduate education. The underlying fear, which was sometimes stated but more often only implied, was that this would result in the department's weakening its commitment to high-quality research and/or would compel individuals to devote more time and attention to their teaching. An unexpected source of active resistance came from a number of senior faculty who were widely recognized for their teaching skill, based on student evaluations, but whose reputations had been built upon being great performers in the classroom while giving traditional lectures. Presenting data on the ineffectiveness of traditional lectures and calling for the shift to more effective pedagogies and different measures of teaching effectiveness can be quite threatening to such individuals.

Finally, a significant source of resistance was the belief in some departments that they could not make any sort of commitment that a transformed course would continue to be taught in that manner, because "we cannot tell the faculty how they should teach." The choice of what was taught and how it was taught in a given course was considered to belong entirely to the individual faculty member teaching the course. Three departments sent in proposals requesting money but made it clear there was no commitment to doing anything beyond asking the faculty if they would like to make changes in their teaching. This individual "ownership" of courses (sometimes even claimed to be a matter of academic freedom) is an aspect of departmental culture that was an ongoing challenge for both SEIs across all departments. It was second only to the formal incentive system as a barrier to change.

The size of the potential grants affected how resistance was handled. The leadership in the department, primarily the chair, had to make a decision about how much time and political capital they would invest in building a consensus—including, possibly, enlisting sufficient support to overwhelm the opposition. Although it varied by department, the chairs at CU seemed less willing to do this than the chairs at UBC, probably because of the smaller grant size. The greater and more visible support from the administration at UBC may also have been a factor.

Evaluation and Funding of Proposals

Once proposals were received, care had to be taken in evaluating them. Then key decisions had to be made regarding the timing and size of grants. It was important to allow for faculty incentives as part of the proposal. Each of these points is discussed below.

Evaluation of Proposals

Because it would take time for departments to figure out the specifics of such a novel effort, the initial funding decisions were made primarily on the basis of how much commitment and general buy-in was indicated. This primarily involved getting a general sense of the department's overall level of commitment, as conveyed in the proposal, and the structures in place for overseeing undergraduate education and its improvement.

Early experience showed that the text of the early proposals per se was not a good indicator, particularly with regard to a department's general sentiments or the functioning of departmental structures. There was often a serious disconnect between the broad commitments expressed in the proposals and what was actually done to fulfill those commitments once funding was provided. As noted above, judgments based on the proposals were more accurate when there were more specifics, such as milestones and timelines and individual faculty names attached to the work to be carried out, and so the requirements for such details were added to the later calls for proposals.

As discussed later in this chapter and in Chapter 6, the organizational structures within a department and the abilities of the people filling the necessary management roles were very large factors in the degree of success of each department's SEI efforts. These turned out to be difficult to evaluate from a proposal alone. Requiring the proposal to explicitly state which individuals would be responsible for filling these roles, and then carrying out a separate evaluation of the commitment and competence of those individuals, proved to be the most accurate means of judging which departments were most likely to be successful.

Timing and Size of Grants

Departments were in very different places initially with regard to both their size and their ability to plan and carry through on a proposal, and so the starting time, duration, and size of the grants were adjusted accordingly. The original design goal was to fund five departments at UBC at a level of $2 million each. The $2 million figure was based on the scale of investment described in the organizational change literature as necessary to bring about major change within an organization, and it was also consistent with estimates of the costs needed to transform all the regular undergraduate courses offered by a large science department. The planned funding level per department at Colorado was $1 million. That was an imprecise decision based on the amount of money available; the value of sending a clear message that while not all proposals would be funded, it was highly probable that a department would be funded if it made a serious effort; and the estimate that there were roughly that number of departments capable of tackling major improvements in teaching.

At CU, four departments were funded in 2006 at about $800,000 each. Although it would have been possible to fund an additional department, it was felt that the remaining proposals did not show a sufficient level of commitment. In the majority of cases, the most serious problem with the proposals was an explicit statement to the effect that "We will invest time and money in transforming these courses for the better, but if any faculty who are teaching them wish to ignore these changes and teach a different way, they will be allowed to do so." These statements were put into the departments' proposals in response to opposition from faculty members. Later, three other departments were funded at a lower level, and with lower expectations as to the extent of the transformation. The seven funded departments included the Department of Molecular, Cellular, and Developmental Biology (MCDB) (2006–2011, extension 2011–2013); Integrative Physiology (2006–2012); Geological Sciences (2006–2011); Chemistry (2006–2011); Physics (2007–2011, extension 2011–2013); Astrophysical and Planetary Sciences (2011–2013); and Ecology and Evolutionary Biology (2011–2015).

Although the CU call for proposals had offered the possibility of somewhat larger grants, upon reviewing the proposed budgets I realized that the departments were unable to find productive ways to spend more than $800,000, often because of the limited number of faculty who were willing to be involved in course transformations.

At UBC, there was more variation in the starting point of the various departments, and so there was a decision to have multiple rounds of proposals and funding. The earth, ocean, and atmospheric sciences (EOAS) department and the UBC biology program, were funded in 2007 for large-scale change. In other departments, pilot projects were funded to sustain the momentum begun with the proposal development process and to encourage them to develop stronger proposals for the second round of funding. Most of these pilot grants targeted individual courses, with the hope that such efforts would lead to more specific and realistic proposals for larger-scale funding. Specific feedback was given as to what was needed to strengthen their proposals—usually this involved making more specific commitments about who would be responsible for doing what when, and developing plans for changes that would be more widespread than first proposed. In most cases, this structure led to more successful large-scale proposals in later years.

Table 3.1. SEI funding levels by UBC department

Department	Funding level ($ M)
Earth, Ocean, and Atmospheric Sciences	1.6
Biology	1.8
Physics and Astronomy	1.7
Computer Science	1.3
Mathematics	1.5
Chemistry	0.7
Statistics	0.3

Similar to CU, departments at UBC seldom came up with credible budgets for spending the full amount that was possible. The departments and their total funding levels are listed in Table 3.1.

The original intent was for the SEI grants to have a five-year duration—sufficient time to transform the courses, but a clear signal that this was a limited-time intervention. As the program began, it became obvious that it would take nearly a year after the funding commitment for serious course transformations to begin. That time was needed to hire and provide at least preliminary training to SESs and to decide on courses to transform and plan what would be done with them. During this ramp-up year, little funding was needed. Based on this, it worked best to operate on a six- or seven-year budget plan: a planning year with little expenditure, five years of full funding and activity, and a final "cleanup" year with low funding, when materials and results are archived and there is a graceful transition to teaching the transformed courses without SEI funding.

In reality, the ramp-up and ramp-down times and levels of activity in a given year varied widely across departments, depending on availability of SESs, faculty teaching assignments and leaves, and other factors. It was best to insist on sustained progress, but to exercise flexibility with regard to detailed schedules.

Allowance for Faculty Incentives as Part of the Proposal

One of the reasons that departments had trouble initially finding ways to spend the full amounts of money that were potentially available was that we discouraged spending substantial funds on direct incentives to faculty to participate in SEI activities, such as reducing teaching loads or buy-outs.

In retrospect, it was a mistake to discourage such direct financial incentives to faculty, and that policy was later changed. As discussed in Chapter 6, the SEI experience demonstrated that that the formal incentive system of the institution provides a very strong disincentive to spend time on teaching. Direct incentives to faculty are necessary to counter this inherent disincentive. We initially believed that the amount of flexible money provided by the SEI to departments would provide sufficient incentives to individual faculty members through perceived indirect benefits, but that was not the case. As noted in Chapters 5 and 6, direct incentives to faculty members, such as a reduced teaching course load for a limited time, or summer salary, worked well when handled properly, but to be effective they required specific agreements in terms of deliverables, timetable, and working arrangements with SESs.

Hiring and Use of Science Education Specialists

All departments began by hiring SESs and, with the exception of one department, all incorporated them as critical components of the course transformation effort. Although the number varied according to funding, at Colorado there were typically two or three SESs per department, while at UBC there were typically three or four per fully funded department. Smaller and correspondingly lower-funded departments had as few as one, although having at least two in a department worked better than having only one.

The SES was a new type of position within an academic department, conceived to specifically fill the needs discussed in Chapter 2 of expertise in teaching and learning in the specific science disciplines. As discussed in detail in Chapter 4, the SES worked collaboratively with the faculty member in this course transformation process, and often established collaborations between faculty members. This reduced the energy and initiative required on the part of the individual faculty member, and hence reduced the barrier to change. Having the faculty member and SES working together to transform a particular course according to the SEI model provided a focus for the work that touched on all aspects of the teaching enterprise. The expectation was that such thoughtfully developed courses could then also be readily reused, making teaching both more effective and more efficient.

There was some background for this particular design. A few years before launching the SEIs, I hired Kathy Perkins and together we carried out transformations of two quite different courses following the approach discussed in the "Course Transformations" section of this chapter. This test confirmed that a new PhD in science with an interest in education could develop the necessary mastery of teaching and learning within a period of several intense months, and could then play a major role in creating and implementing courses close to the ideal described. The materials for these courses were then archived and were subsequently passed along and reused with little change over several years through multiple instructors as a matter of tradition (rather than as a result of formal departmental oversight). Both of these courses used large amounts of active learning.

All aspects of the SES work, including hiring, training, supervision, and the jobs they carry out, are discussed in Chapter 4.

Departmental Organization for Managing SEI Efforts

It was challenging for most departments to organize and operate the SEI efforts. No suitable organizational structure existed for such work, nor was there local expertise as to how to make such an enterprise successful. Existing structures, such as the undergraduate course committee (often called the curriculum committee), are inherently reactive and so were ill suited to the task. Thus, over time, I put more requirements in place for structures to be established before funding was provided, based on structures that had worked well in successful departments.

In well-functioning departments, there was a SEI department director appointed to oversee the SEI activities. This person had clear authority, including hiring and supervising the SESs. The SEI department director's duties included:

- Overseeing the hiring of the SESs
- Determining how the SEI money would be spent
- Supervising the SESs (that is, SESs reported to the department director)
- Establishing the job expectations and requirements, including deciding which courses and faculty members the SESs would work with
- Establishing the working arrangements between SESs and faculty members
- Intervening when problems arose (such as with faculty members not fulfilling commitments with regard to collaborating with the SES)

- Meeting regularly with the department chair to report on activities (the chair arranged for regular reports to the department about the SEI work and accomplishments in faculty meetings and other venues)
- Arranging any faculty incentives supported by SEI funds, usually in conjunction with the department chair
- Under ideal circumstances, having some input into teaching assignments (but in no case did the SEI department directors have as much influence in this area as they would have liked)

The structure by itself was not entirely sufficient; how well the department directors functioned and were supported within the department made a large difference in how successful the SEI efforts were. When a department was funded (or under serious consideration for funding), SEI Central would sit down with the department chair and work out exactly how this necessary organizational structure would be established within the department and who would be the director. In one department, funding did not go through when it became apparent that no one in the department was willing to serve as the SEI director. This was a sign that the department did not see this as a sufficient priority.

When the department did not establish the chain of command as described above, typically the SESs would view SEI Central as their supervisor, and would come to SEI Central when they experienced problems working with faculty. This was a bad situation, because it was difficult for SEI Central to deal with problems within departments, and trying to address such problems made the SEI appear as a program that was being pushed on the department, rather than something the department was responsible for and invested in.

Course Transformations

This section describes the implementation of an extensive course transformation involving an SES and faculty. The heart of the SEI was the process of course transformation, in which an SES worked with faculty members to transform courses, and simultaneously the teaching methods of the faculty, according to the SEI principles. Typically, one SES would work with a sequence of faculty members to transform a sequence of courses. The details of scheduling and sequence varied substantially with department and courses, but a typical situation was an SES working simultaneously on three courses: the pre-transformation planning stage for one course, the full transformation of an ongoing course, and follow-up, refinement, data collection, and analysis of a second iteration of a transformed course. The bulk of their time would be spent on the full-transformation course. For large and complex courses, it was not unusual for the full transformation stage to require more than one iteration of the course, with new elements and activities phased in and/or modified over multiple offerings of the course. The SES collaborated heavily with the faculty member during each step of the course transformation, taking on many of the labor-intensive duties that teaching faculty did not have the available time or expertise to attend to. In Chapter 4 I go into more detail about how the specific and rather unique elements of the job of SES contributed to the course transformation.

In a few cases, a small working group of faculty would come together to oversee the transformation of a course. At times that model worked well, with useful contributions from and interactions between multiple faculty members, and the resulting course goals and design gaining elevated stature within the department. Frequently, it was less successful, with only one or two faculty feeling it was worth their time to be involved. In some other cases, working groups were organized but functioned badly because one or a few faculty members in the working group were serious impediments to accomplishing anything, either through active opposition or simply because of their failure to fulfill agreed-upon responsibilities. In most cases, the SES still worked quite productively with a single faculty member to carry out the desired course transformation.

Typical Course Development Cycle

Below I discuss the ideal process for developing a course. However, the degree to which this cycle was followed varied considerably, primarily affected by the desires of the faculty involved and to some extent the departmental management of SEI efforts. It was not unusual for the order of steps to be changed or some steps entirely left out. Also, in some cases the SES worked more as a consultant to many faculty members in the department in regard to making incremental changes, rather than focusing on transforming a specific course(s). In that role the SES would provide advice on instructional activities that a faculty member decided to add to a course they were teaching. The model of full course transformation was preferable, as it seemed to generally result in a higher-quality product, but there was flexibility to pursue all possible opportunities for adoption of improved instructional practices. In most cases, over time the SES took on both roles, working with individual faculty to transform specific courses while serving as a consultant to much of the department.

Outlining the Project Scope

An essential first step was for the SES and faculty member(s) involved to agree on what the project was to accomplish, and the respective responsibilities and expectations. In practice, few course transformations proceeded by working through the eight steps in Table 3.2 in a smooth orderly manner, and various different weightings were given to the three elements of establishing goals, assessments, and teaching methods. Although there was a large amount of variation in the process, often the SES would start by discussing with the faculty member(s) any issues or problems involving the course in question. The SES would then investigate and propose possible directions and activities to address the most salient problems, and then, as the relationship developed, build from there to try to address all seven steps of the course transformation in whatever order the faculty member found preferable. In particular, starting with learning goals turned out to be a problem with many faculty members; it was just too difficult for them. They had an easier time starting with what student difficulties they wanted to address, what sorts of activities and assessments they wanted to use, and what material to cover and why. After becoming immersed in those issues and establishing greater interaction with students through the use of more interactive teaching methods, they then had an easier time articulating learning goals for the course.

However, when it was possible to work through the eight steps in an orderly manner, the results were usually best, so the implementation is discussed in that order.

Developing Learning Goals

Learning goals define what a student should be able to do as a result of learning the material. Both course-level learning goals and topic-level learning goals were typically developed. Approximately five to ten

Table 3.2. Central features of course transformation planning

Steps	Description	Tasks
Project scope	What do we want to accomplish?	Meetings—establish deliverables and timelines
Course- and topic-level learning goals	What do we want students to learn (for example, content, skills, habits of mind, attitudes)?	Meetings, create, review
Document student thinking	How do students think about the material of the course, and what do they know coming in?	• Do literature review • Observe course before and after transformation • Interview students
Teaching methods	How will we help them learn the material?	• Create course materials and activities that target learning goals, consistent with research • Select teaching practices and course structures best suited to material, constraints, and faculty desires
Assessment	How do we know if students achieved the learning goals?	• Exams, conceptual assessments, homework • Pre-/post-course surveys • Student interviews
Materials archived	How will others find/use what we've done?	Organize materials locally and online
Plan for sustainability	How to support adoption and/or adaptation of course materials and methods by others?	Interact with faculty and administrators prior to and following transformation; implement support and transition strategies, such as co-teaching

course-level learning goals were created, which were broad and not necessarily related to particular course content (for example, "students should be able to simplify real-world problems in terms of basic physics concepts"). For each topic, several learning goals were developed that were more specific and represented a concrete step toward achieving a course-level goal (such as "students should be able to construct a free-body diagram depicting the forces on an object").

Learning goals are more specific than a listing of topics. All learning goals needed to be operationalized so that their accomplishment (or not) was *measurable.* It was very common for goals to be proposed that were too general or vague so that it was unclear how students would demonstrate that they had successfully achieved that goal. For instance, the initial attempt to produce learning goals commonly included "Students should understand . . . [various topics]." Such goals would then need to be rewritten, since two faculty members could have very different ideas of what "understand" means in the context of the course. These goals were rewritten in terms of what students would be able to do if they understood the topic or concept at the desired level.[3]

Assessing Student Thinking and Learning

The SEI process for assessing student learning typically began by soliciting input from faculty who had previously taught the course and faculty who had taught students in subsequent courses, in order to identify student weaknesses. Next came consultation of the discipline-based education research related to the course material and an examination of student performance on exams, both the standard course exams and, where available, validated third-party tests covering the material. Finally, there were student interviews, both formal and informal, on the course material. Frequently, the course exams were modified as a result of this process to better target the goals that were arrived at. A detailed description of investigation of student difficulties in the CU Physics Department that was part of an independently supported research project (and hence was more extensive than many other SEI efforts) is described in work by Chasteen et al.[4]

Creating Course Materials and Implementing the Instruction

There are many models of how to create course materials, but the most important thing is that the course be aligned with the established learning goals and that the materials provide practice and guiding feedback to the students, informed by known student learning difficulties. Specific strategies and teaching techniques of the sort that were used are discussed in Chapter 2 and the references given there. The SESs were trained in the use of these techniques, as well as with the education research literature, so that they would be able to provide insight and guidance on possible teaching options. They would collaborate with the faculty member to apply this knowledge to the specific material and learning goals to be covered. Often there were a number of research-based teaching methods that could be used in a given context, and it usually wasn't obvious whether one specific method would be more effective than the others. The choice of which method to use was often determined by the faculty member's interest in or comfort with a specific method.

While each individual case was different, a common path for an instructor was to proceed incrementally, starting with modest changes and then building on those changes. The change to standard lecturing that was usually easiest to start with was introducing questions to students into the middle of lecture, followed by student-student discussion. Usually this was in the form of "clicker questions" and "peer instruction" or "think-pair-share."[5] These were good initial steps in transforming instruction, as they involved relatively small changes by the instructor, but they provided opportunities for greater interaction between instructor and students, which would typically result in instructors making further changes as they better understood student thinking and saw improved student engagement. Other relatively readily adopted new teaching methods included in-class worksheets, placing TAs in large lectures to facilitate group discussions, two-stage exams, concept mapping, learning to circulate among students and listen to conversations about activities, and providing learning goals to students before and during class. SEI Central worked with the SESs to develop short (one- or two-page) guides for faculty on optimal implementation of these and other commonly adopted teaching methods. These were posted on the CWSEI website (www.cwsei.ubc.ca/resources/instructor_guidance.htm), and many are included in Appendix 1.

Sometimes these new teaching methods would first be demonstrated in the course by the SES while the faculty member observed. Somewhat more frequently, the SES would only provide coaching and guidance to the faculty member as he or she implemented the methods. The SES would typically observe most classes during the first implementation of a transformed course, providing assistance as needed and feedback to the instructor after each class. As there are countless ways to do most any teaching method incorrectly, a critical role of the SES was to know the principles of learning that lay behind specific techniques and the specific elements of implementation that could help and hinder the effectiveness of that technique, and then pass those along to the instructor.

This knowledge and its effective transmission were an important part of the SES training.

Assessing Course Outcomes

Another aspect of scholarly course transformation is the use of assessment data to allow for reflection and iterative improvement upon the transformed course. As discussed in Chapter 5, the type and extent of outcome assessments varied wildly and was generally less than desired. Among the choices were common or similar exam questions or other student work, such as clicker questions or homework, compared across years; instructor-independent measures, such as concept inventories, used to test students on content mastery; diagnostics and performance in subsequent courses; ability to answer more difficult exam questions than in previous years; student interviews; and classroom observations, usually using the Classroom Observation Protocol for Undergraduate STEM (COPUS).[6] In too many cases little assessment of the course was carried out beyond the instructor's impressions. Although the data were limited and varied in type, in virtually all cases where data were available, they showed improved results in the transformed cases.

Dealing with Faculty Teaching Rotation

There are very different policies about the rotation of faculty through courses, depending on both the department and the level of the course. It was difficult and inefficient to transform courses in which different faculty members rotated through too frequently, and also difficult when there was too little rotation. In the case of frequent rotations, a faculty member might teach a particular course intermittently a few times over a period of several years or teach the course for a couple of years and then move on to an entirely different course. In these cases, it was very difficult to work with an individual faculty member on transformation of that course, because (1) the faculty member who helped develop the transformed course materials might not teach the course again soon, and (2) an individual faculty member might not have a great deal of incentive to invest the kind of time required to transform a course because he or she would not benefit from the effort.

Too frequent faculty rotation remained a nagging problem for the SEI.[7] However, a few approaches have been helpful in reducing the problem. I pressed the departments to have a faculty member teach the transformed course multiple (typically 3 to 4) times, and/or for faculty who are experienced with active learning to teach the course subsequent to the transformation. I encouraged the departments to partner the SES with multiple faculty members in succession in the course, both to transform the course and support new faculty members in teaching it. Finally, I insisted on the departments setting expectations that the SESs would create well-organized, easy-to-use course archives and give faculty members new to the course an introduction to this archive. When there was little or no rotation of faculty through courses, each course was essentially seen to be "owned" by a faculty member, with the teaching and topics entirely a matter of that person's individual choice. As discussed in the "Barriers to Change" section of Chapter 6, this made changing the teaching of such courses quite difficult.

Co-Teaching

An alternate approach to preserving the benefits of a transformed course and transforming the teaching of faculty members was to have another faculty member co-teach a transformed course with the instructor who had carried out the transformation. These arrangements worked best when the two instructors truly worked as a team (rather than trading off course duties), including planning the course together, coming to most classes, and jointly developing exams. A variation on this model was to partner an SES as a co-teacher with a faculty member who was new (or relatively new) to the course.

While only a select group of faculty participated in such co-teaching programs, they were almost all highly successful. New faculty members reported that they spent less time on teaching preparation than many of their junior colleagues, and they quickly became highly effective teachers—in some cases, among the best in the department. Established faculty members who took advantage of co-teaching reported that they greatly enjoyed the experience and learned a great deal from observing their colleagues teach—something that is usually rare in a department. As a result of such successes, and to support sustainability of better teaching methods, external funding has been obtained at UBC to establish a long-term program of such co-teaching in some departments for the purposes of faculty development.

SEI Central Oversight

SEI Central served three basic purposes. First, it was an engaged funding agent participating in the development of proposals and making funding decisions. Second, it provided oversight of the departmental activities and gave feedback, particularly on how to improve the results. Third, it served as a trainer of SESs. That training also included providing substantial individual guidance to the SESs on both pedagogical issues and on effectively working with faculty and dealing with difficulties in the department.

Administrative Role

SEI Central played an important administrative role, taking on responsibilities that would not have been supported by departmental structures. As such, SEI Central required some funding, employing a director and/or associate director and administrative staff. At UBC this consisted of two or three FTEs in the earlier years, and one FTE in the final years.

SEI Central responsibilities included:

- Soliciting and reviewing proposals
- Administration and oversight of funding and budgets
- Advising on hiring of SESs
- Training of SESs
- Assisting SESs with design and analysis of interventions and assessment, and with writing up and publishing of education research papers
- Support of SES community (planning of regular meetings, providing information, and participating in discussions on SES forums)
- Monthly meetings with each departmental SEI team (the SEI department director and SESs)
- Quarterly meetings with the group of SEI department directors and the associate dean
- Soliciting and providing feedback on annual reports from departments
- Running an annual SEI mini-conference with presentations and posters
- Website maintenance, including course materials archive
- Collaboration with other institutions

Training Science Education Specialists

As no new SES had the preparation needed to serve in the role effectively, and neither the departments nor the universities' centers for teaching and learning had the knowledge or capacity to train the SES, this training was one of the most important jobs of SEI Central. The details of the training are discussed in Chapter 4.

Annual SEI Mini-Conference Events

Annual half- to full-day gatherings served to celebrate and show off SEI accomplishments, as well as support a community engaged in educational work. These events featured talks by a mixture of faculty and SESs, poster sessions, lunch, and workshops on various aspects of science teaching. These were intended to serve as an important dissemination and recruitment tool, attracting many faculty from both participating and nonparticipating departments to learn about the SEI activities within their departments and elsewhere. Unfortunately, few faculty attended these events who were not already directly involved in SEI activities. However, it did serve as a good way for participating faculty, SESs, and graduate students from across the SEI departments to learn what others were doing and generate a sense of being part of a large and vibrant program. At UBC, the dean and associate deans were always prominent attendees, demonstrating their support for the SEI. These annual events also attracted a number of visitors from other institutions, coming to learn more about the SEI. The poster sessions proved to be particularly lively and well attended. Sample materials and an agenda from such a conference are available at www.cwsei.ubc.ca/EOYevent2014.html.

Central Resources and Website

At UBC, SEI Central took responsibility for developing an extensive website providing resources for faculty, SESs, and the outside world (www.cwsei.ubc.ca). This includes detailed information on how to carry out course transformations, specific topics in teaching and learning, SEI publications and presentations, instructional videos, videos showing instructors and students in transformed courses, and other material. A small SEI library is also maintained, with about twenty particularly valuable reference books available to SESs and faculty working with them.

To facilitate the transfer of courses, a fairly elaborate online system was constructed that allowed materials to be easily archived and accessed. This was more difficult and expensive than anticipated for a variety of reasons. While an extensive survey of user needs was administered when designing the system, later on when it came time to use it, users wanted different capabilities, and different departments and individuals had strong opinions about specific details. Materials for some transformed courses were posted on the website, but it was disappointing that full sets of materials were posted for only a small fraction of the transformed courses. One of the reasons that many departments have been exploring the hiring of someone to an SES position on a permanent basis is to facilitate archiving and dissemination of materials, as they have found it very difficult to get faculty members to do this.

Program Oversight by SEI Central

As noted above, closer oversight of departments was needed than initially expected. Progress was monitored through a combination of inputs. First, written summaries from the SESs were required (every two weeks in the early years, monthly in later years) on what they had accomplished. Frequent individual meetings were also instituted with SESs, particularly when they were encountering problems.

Typical problems encountered by SESs included difficulties working with a faculty member, figuring out how to juggle multiple priorities, and carrying out research (for example, trying to define research questions or produce publications). Problems of the first type were the most common and most serious. In the early years, it was quite routine for a department to assign the SES to work on a course, but the faculty member teaching that course had no interest in being involved, and no one in the department could or would intervene. Alternatively, the faculty member with whom the SES was to work was nominally agreeable to the collaboration, but then in practice would not cooperate. For example, the faculty member might always be too busy to meet with the SES or provide them with the course materials, or would only send lectures or activities to the SES for review and feedback just before class. Occasionally SESs would be working on a course in which multiple instructors were involved who had fundamentally different goals for the course, which meant that the SES was caught between opposing views and unable to make progress. SEI Central played an important role in helping to advise SESs on such challenges, and over time, we learned how to manage and avoid problems like these. The main improvements were to get departments, primarily through the SEI department director, to be more proactive in avoiding such problems and providing support and guidance to the SES early as problems started to arise. (See Chapter 5.)

When things were working well, each report from the SES would provide a new list of changes made in courses or new assessments of student learning and difficulties. Thus it was very easy to see progress, and correspondingly easy to see when there were problems. Although poor progress was most commonly due to SESs being assigned to work with faculty who were uncooperative or did not understand the expectations, there were times when the problem lay with the SES. When we became aware of a potential problem, we would meet with the SEI department director to better understand the source of the problem and figure out steps to deal with it. Such issues were much easier to deal with in departments where there was a department director who clearly understood that this was part of the job and had the authority to address the problem, whatever the source.

Monthly meetings with departmental SEI teams (the SEI department director, all SESs, and others, including graduate students and undergraduates hired temporarily) were very useful to review progress and plans. These were the primary opportunities for SEI Central to provide feedback to the department on its progress and to provide input on proposed directions.

There were also quarterly meetings with the SEI department directors and, at UBC only, the associate dean and occasionally other members of the university administration. These meetings allowed departments to share various practices and approaches, such as the most successful ways of incentivizing faculty and ensuring good working relationships between SES and faculty.

When there were serious concerns about lack of progress in a department or special problems, we would meet with the chair. Usually, but not always, this was at our request.

The most extreme element of oversight was terminating funding for a department. In the two cases where it was first proposed and then carried out, this served more to avoid wasting money than to bring about changed behavior in response to the threat of termination. This was an indication of a flaw in the initial SEI implementation design. More funding should have been allocated to direct incentives to the faculty, so that the threat of losing that funding would have been more of a concern. Instead, where termination became a possibility, essentially all of the department's SEI funds were going into SESs rather than any other departmental support. Because the lack of progress stemmed from the unwillingness of the faculty to work on changing teaching, with or without the assistance of an SES, the loss of funds, and hence the loss of the SESs, was unimportant to them. This lack of priority placed on maintaining the SES was clear when I talked with the respective department chairs about the possibility of termination.

When termination of funding became necessary, we worked out with the SESs the timing and conditions for their graceful exit, and simply did not provide funding to the departments for their replacement. In one of the cases where funding was terminated (the UBC biology program), there was a subsequent change in department chairs and a major restructuring of the way the undergraduate program was run and overseen. With the new organizational structure and good people in positions of authority in that structure, the funding was then restarted, and there was good subsequent progress.

Challenges with Data Collection

As described in Chapter 1, my vision of the SEI was that it would be a step toward a data-driven education system, where educational data was routinely collected and used to improve outcomes. I had expected data to be regularly collected on student outcomes from courses (learning, attitudes about learning, and interest in subject) and instructional practices. I had also hoped to obtain data during the SEI on shifts in the departmental cultures, particularly the attitudes about teaching represented in those cultures.

The implementation of the SEI revealed a number of intrinsic challenges with collecting data on each of these outcomes. A substantial amount of data is presented in Chapter 5, but there was a large variation in the quality and quantity of data across the different outcome measures and institutions. Most of these difficulties were unexpected but in retrospect are understandable, as they are inherent in the structure and incentive systems of the institutions.

Challenges with Collecting Data on Student Learning and Attitudes

Originally, I thought this would be the most important outcome and straightforward to measure. In practice, it turned out to be quite difficult to track.

We were able to systematically collect these data in only a small fraction of the transformed courses in any department. The basic problem is the disincentive for individual faculty members and departments to collect data on student learning and other outcomes, particularly baseline data for courses and programs before they are transformed. This was not standard practice in any department, it takes work to collect these data, there is no benefit to the faculty member for doing it, and the results may reflect poorly on the faculty member and possibly the department as a whole. However, there are also a number of more specific issues encountered when trying to get such data, as discussed below.

First, one needs to get baseline data before a course is changed in order to determine the impact of any instructional changes, and there are numerous challenges in getting such baseline data. Generally, nothing exists except instructor-dependent measures, such as student performance on test questions created by the instructor. These tests tend to be highly idiosyncratic and usually of questionable validity, as instructors have no training in creating good tests and seldom get any feedback on the quality of their questions. So the instructor exam data are often unsuitable to serve as a baseline.

In a small number of cases where the course topics have been the subject of discipline-based education research, there are independent tests that have been developed to target the mastery of particular topics covered in a course. When such tests exist, they are very useful to use on a pre- and post-course basis to measure learning, but typically there is considerable instructor opposition to the use of such tests in untransformed courses, making it hard to get baseline data. Those instructors were usually quite resistant to allow such outside measures of their students' learning to be carried out, either because they felt threatened by the process or because they believed it to be a poor use of class time.

Even if the instructor in a course was supporting the future transformation, it could be difficult or impossible to get meaningful baseline data. Usually a good assessment, such as what would be produced by an SES in consultation with course instructors and SEI Central, is developed only while the course is being transformed, so there is no longer a traditionally taught course available to use as a comparison. Instructors who are interested in changing a course do not want to continue teaching the course without change for a year just to get baseline data. Finally, as instructors learn new and better ways to teach, they usually end up modifying (and improving) their learning objectives and test questions. As a result, data they may have on student performance prior to the transformation, such as answers to exam questions, often no longer apply, because instructors no longer feel that such questions are appropriate to use after the transformation. In spite of these challenges, for roughly 10 percent of the transformed courses at the two institutions there have been common (or quite similar) good exam problems or instructor independent measures that are given year after year and can be used for comparison.

Another issue that arises in interpreting data provided by graded exams in courses involves departmental expectations around grading and failure rates. In several cases in which students' performance on similar exam questions improved as the result of changes in teaching methods, the instructors (often in response to pressure from their department) increased the difficulty of the exams, to have the grade distribution match departmental norms (for example, a B average). Thus, measures such as student grades and failure rates may be kept constant or may vary idiosyncratically with the instructor, independent of the student performance in any objective sense. Thus, I learned that student course grades seldom provided a meaningful comparison of the amount of student learning achieved with different teaching methods.

I was also interested in the impact of the transformed courses on students' attitudes as another measure of success. Did the course increase or decrease students' interest in the subject and/or their desire to pursue a career in the discipline? How did it impact their views about learning the subject and the best ways for them to learn? There are validated survey instruments suitable for measuring some of these attitudes,[8] and in other cases non-validated questions (such as "Did this course increase your interest in taking another course in the discipline?") seem adequate. However, collecting such data from courses still proved difficult. If students are given such a survey during class and asked to fill it out, they usually will comply. Yet, few instructors were willing to use class time for this purpose. Most students are unwilling to take the time to complete such surveys outside of class, unless they are given a small amount of course credit for doing so, but the majority of instructors are not willing to allow course credit for this. As a result, although there are some encouraging hints that course transformations improve student attitudes toward the subject and learning, there were few courses in which the survey completion rates were high enough to provide confident results.

Confusion over Human Subjects Research Rules

A unique barrier to collecting data on educational outcomes came from some administrators, faculty, and lawyers at these two institutions who misunderstood the rules regarding human subjects research at universities. (The rules in Canada and the United States were basically the same.) I discuss this as the same issue may well be a problem for readers who wish to collect similar data at their own institutions. I encountered this at both institutions, but with some considerable investment of time and effort eventually overcame this particular problem at UBC.

The idea that rigorous assessment of the learning in courses, and hence the effectiveness of teaching, could be carried out and disseminated without being treated as human subjects research was not an accepted concept on either campus. Although there is a specific exemption within the federal human subjects research rules to cover evaluation of institutional quality for organizations such as educational institutions, few people at universities are aware of this exemption because such evaluations are so seldom done. Thus their first reaction was to treat any effort to collect data on student performance as falling within the category of individual curiosity-driven human subjects research. When treated as curiosity-driven research, such assessment of learning was thought to be subject to extensive paperwork, institutional review board (IRB) approval, and collection of signed student consent forms. This led to the bizarre initial stance of the respective IRBs saying it was completely up to the faculty what teaching methods they inflicted on students in classes, but if the SEI program wanted to measure what effect these practices had on students, that data collection would require a lengthy review and would be subject to written preapproval by students in order to avoid harming the students. The signed consent forms are a major burden, since they take a lot of time to collect and process, and students often fail to turn in the necessary forms and/or are understandably concerned and suspicious when consent forms filled with all the required legalese are presented to them.

Dealing with this issue of misinterpretation of the rules regarding human subjects and institutional research for quality improvement required a large investment of time on my part, including becoming the campus expert on the wording of the legal statutes and regulatory language on both human subjects research and student privacy statutes. The common standard set by many IRBs, which was initially invoked by IRBs at both CU and UBC, was that any study producing data that might result in any form of publication is curiosity-driven human subjects research and requires IRB review and approval. Because it was important for both internal credibility and long-term career success that the SESs be able to publish their work on documented instructional improvements, this standard posed a particular burden. There was a lengthy negotiation at UBC to deal with these issues, resulting in a clear delineation of institutional research versus traditional academic ("curiosity-driven") research, and the establishment by the University Counsel of institutional policies governing the conduct of the SESs and the faculty in SEI departments. Among other things, this led to a change in the UBC rule that anything involving publication required IRB approval.

I found that it was still important to have clear guidelines for SESs and faculty working on SEI-supported projects as to which studies and data collection did and did not need to go through IRB review. It was fairly easy for SESs or faculty to get so interested in an education research question that they would forget the special responsibility that goes with carrying out research in a real class that students are taking for their education. SEI Central or a suitable administrative person in the departments needed to briefly review all proposed studies to decide which might need IRB review and approval and which would not, and if the study would be raising any ethical issues.

The most common ethical concern that arose involved the establishment of a control group that was to receive traditional lecture instruction. I set the policy that it was unethical, and hence not allowed, for SEI involved personnel to establish such a control group on the grounds that they (or any reasonably informed person) had good reason to believe that students in the control group would be disadvantaged. This was true even if the control group was to receive instruction that is consistent with common teaching practice and many research studies. On several occasions, SEI Central pointed out that proposed studies by SESs and/or faculty working on SEI activities that would have set up such a control group were not appropriate. On the other hand, when there was a faculty member who was going to teach using traditional lectures as part of his or her regular course instruction over which the SEI had no control, irregardless if a study was to be carried out, then it was ethical for SEI personnel to use that class as a control group, as students would not be disadvantaged due to the actions of the SEI-supported people.

Table 3.3. A set of guidelines regarding human research subjects' questions

Type of research	Definition	Data collection and publication guidelines
Studies of effectiveness of normal instructional practices using existing or routinely gathered data	Practices that are commonly being used in university classrooms at the present time	Data collection (including subsequent publication where warranted) needs only minimal review by any person at the university in a relevant administrative role (SEI director or associate director, department chair, or SEI department director)
Studies of large-scale changes in courses using research-based practices and assessments in use elsewhere	Significant changes are being made in the teaching of a course, particularly using methods that are not in such common use, and so the potential impacts are larger	Studies of these sorts of changes are subject to the same type of approval as above, but there is closer examination of the potential benefits and risks to students in both intervention and control groups, and steps to reduce harm. For example, arrangements might be put in place to readjust student grades in one group if it turns out that the other approach is superior, resulting in significantly higher performance for some fraction of the student population on a common exam.
Non-course-specific educational research	Involves a selected subset of students in activities such as individual interviews about educational experiences transcending a particular course	For the sake of expediency, we went through the standard IRB review process to get a broad approval for such activities. These were still considered "exempt" by IRB standards (not requiring full-scale review), as they involved minimal risk but do require consent forms. Much of this work could have been categorized as institutional research, according to the definition provided by the UBC University Counsel.

For illustrative purposes, Table 3.3 lists the basic guidelines that I worked out with the UBC university counsel. (These are in my wording, which removes the legalese but makes them easier to understand.)

Challenges with Collecting Data on Shifts in Departmental Cultures

Data on the broader impacts of departmental attitudes and culture are quite limited. The primary difficulty with collecting good data on this topic is simply the expense. Done optimally, it requires an objective independent evaluator to carry out extensive interviews, surveys, and observations of faculty and departmental staff as they conduct their instructional work and departmental business. While it seemed ethical to use institutional funds—which were in very short supply—to support the SEI improvements in teaching that would benefit students and faculty at those two institutions, it did not seem appropriate to use those funds for collecting data on shifts in attitudes and cultures of the departments. That data would primarily benefit those in the outside world who might want to use the results of the SEI to launch and guide their own similar efforts, but the results would come too late to make an impact at UBC or CU. Repeated attempts to secure external funding to support such studies of institutional change were unsuccessful.

Nevertheless, to the extent that it was possible to do so with minimal cost to the SEI, some data were collected on departmental attitudes. SEI Central sampled those attitudes as well as it could through the review of SES reports and departmental discussions, interviews with SESs, faculty, and department administrators, and some surveys. On a limited basis, we also had external researchers come in and sample faculty and/or SES views. The extent of the data is better from UBC, because we put greater emphasis on regular meetings and written reports from the SESs and on more frequent and formal meetings with departments. In essentially all cases, the different sources of input were quite consistent, but for some CU departments there was less consistency and hence greater uncertainty as to the general attitudes and response within a department.

Challenges with Collecting Data on Teaching Practices

The collection of data on the teaching practices used in courses was not done on any scale at either institution before the SEI, and getting such data also encountered challenges. However, these challenges were less of a problem than those discussed above. First, there was little need for collection of baseline data to determine changes, because it was so unusual for a faculty member to use anything besides traditional lecture and recitation practices. Typically, if anyone was incorporating research-based methods, it was a well-known anomaly in the department. Second, the adoption of new teaching practices was so central to the SEI activities that by monitoring the activities of the SESs, the courses they were transforming, and the faculty they were working with, we could get reasonable data on the teaching methods used in various courses and the changes that had been made. As noted earlier, the more regular and detailed reporting by the SESs and departments at UBC compared to CU provided us with more complete and reliable UBC data. The annual departmental SEI reports required at UBC as to courses transformed, changes made in those courses, and faculty involved in those changes, were particularly useful.

Substantial effort was also put into developing tools that could be used by departments for routinely monitoring teaching practices. The COPUS (Classroom Observation Protocol for Undergraduate STEM)[9] was an easily used observation guide for characterizing how the students and the instructors were spending their time during class. Various SEI departments are using the COPUS in characterizing and offering guidance to their instructors, and it is now also being used widely outside UBC and CU. Sarah Gilbert and I, in collaboration with numerous SESs, also developed the Teaching Practices Inventory (TPI).[10] The TPI is a survey usually filled out by instructors that takes about ten minutes and provides a detailed characterization of all aspects of how a course is taught. It provides extensive and complete data on the teaching practices used in a course, so in any course for which it was used, we have detailed data on the teaching practices.

However, there were institutional challenges in getting faculty to fill out the TPI. Although we hoped that many departments would make this part of their regular annual reports by faculty because it provided so much more information about teaching than they had been collecting, this did not happen. At UBC there was sufficient financial leverage to get most departments to put in a reasonable effort to get their faculty to complete the survey on a one-time basis, and one department (EOAS) had most of its faculty complete it at both the beginning and near the end of their SEI funding. At CU, after negotiation, the dean urged the science department chairs to ask their faculty to fill out the survey. Only a few chairs did so, and in those departments only a few faculty completed it, so no useful information was provided. Data on teaching practices at CU came primarily from SES reports and surveys and interviews SEI Central did with the faculty. These were less detailed and complete than the data from UBC.

These difficulties with the collection of data reveal how large a shift will be required in institutional and departmental cultures before routine data-driven educational improvement becomes the norm. The data that was collected on student outcomes, departmental culture, and teaching practices, and how these varied across the SEI departments, is presented in Chapter 5.

Common Obstacles and Desirable Elements for Successful Implementation

There was a complex range of factors acting at various levels to both enhance and inhibit the success of the departmental SEI efforts. Although every department had its own unique characteristics, there are many things that were consistent across departments, both in what worked and what caused things to fail. These are discussed in Chapter 6. Here I briefly list the most prominent obstacles and elements of successful departments encountered during SEI implementation efforts.

The first common obstacle was a lack of faculty commitment to the proposed work. When it came time for individual faculty to do the work called for in the departmental proposal, some of them refused. The willingness of new chairs to live up to the commitments of previous chairs was also a problem at CU.

Second, individual faculty "ownership" of courses was a common issue. The belief that no one else in a department could or should tell an instructor what or how to teach in a course assigned to that person was an ongoing challenge to SEI efforts. The strength of this belief varied across departments for no apparent reason other than tradition. On many occasions, department chairs and SEI department directors appealed to me to get faculty members to change how they were teaching, indicating that the department recognized that what a faculty member was doing was relatively ineffective but did not see itself as having the authority to tell the offending individual to change.

Another common obstacle can be put under the heading of "thoughtless teaching assignments." Some departments had a tradition of making last-minute, haphazard teaching assignments. This was a major problem for the SEI course transformations, which required consistent planning and implementation over several semesters. Temporary sessional instructors who were hired at the last minute to teach a course for a single semester were a particularly serious problem.

As already discussed, an obstacle consistently encountered was the lack of departmental structures to oversee educational innovation, such as the SEI.

A final unanticipated obstacle was the existence of courses taught in multiple sections by multiple instructors. As discussed in Chapter 6, this is a complex issue that has many local variations. Surprisingly, most such courses were historically operated in such a way that the individual instructors, many of which were long-term non-tenure-track instructional faculty, were largely free to do what they wanted with very little oversight or coordination. In these cases, transforming the courses proved to be quite difficult. As many departments recognized there were problems with these courses, they assigned SESs to work on improving the courses, but without the agreement of the instructors. As a result, considerable SEI funding was spent on these courses with modest results.

FOUR

Science Education Specialists: Agents of Change

A CRITICAL ELEMENT of the SEI was the use of science education specialists (SESs) that were embedded in the departments. These specialists were experts in the discipline with expertise in teaching the discipline using the most effective research-based methods and principles. The SESs worked collaboratively with individual faculty members to change how courses were taught, and to enhance the teaching expertise of the faculty member in the process. Their work on course transformation focused on three key questions: What *should* students learn? What *are* students learning? and What instructional practices will improve student learning? The SESs played a vital role in the SEI change process and were responsible for much of the success of the SEI, but there was little precedent for such change agents. The SEI largely invented this position and figured out how to make it effective in improving teaching. In this chapter, I describe the SES position, and how SESs were hired, trained, and worked with faculty to improve teaching.

What is an SES? These professionals, called "science teaching fellows" (STFs) at CU and "science teaching and learning fellows" (STLFs) at UBC, offer an unusual combination of expertise in their discipline and knowledge of relevant teaching methods and research on learning. A typical SES was a recent PhD in the relevant science discipline who was keen to improve teaching, and to varying degrees interested in education research. However, there were also a number of excellent SESs with different backgrounds.

These included people who had been frequent instructors in the department on short-term contracts, an emeritus faculty member, master's degree holders, and graduate students (for limited periods). Whatever their background and whatever the process by which they were selected, all effective SESs combined thorough disciplinary knowledge with good interpersonal skills and a strong interest in teaching.

Since most individuals hired as SESs had limited prior experience with research and research-based teaching methods, new SESs attended a semester-long training program run by SEI Central. SESs also attended ongoing meetings to further develop their skills and to generate a cross-departmental community and learning opportunities.

The primary job of the SES was to collaborate with individual or small groups of faculty to implement course transformation, helping faculty members increase their knowledge of relevant teaching and learning research and supporting the introduction of evidence-based educational practices and measurements of learning. It was important for SESs to be partners and gentle coaches for faculty—and not to be treated as glorified teaching assistants (TAs) who merely develop instructional materials.

The most successful SESs were those who were viewed (and viewed themselves) as *departmental resources*, and therefore continually worked to enhance both their scholarly expertise about teaching and their productive relationships with many faculty members. These SESs acted in a variety of capacities:

- Supporting specific course transformation efforts, as described below, including documenting achievements and archiving materials
- Being a consultant for general faculty questions on effective teaching, or small teaching projects (that is, not a full course transformation)
- Running workshops for faculty and/or TAs on various teaching methods, or bringing in outside experts to do so
- Engaging the department by running seminars or brown-bag luncheon discussions on teaching issues, creating newsletters, and actively seeking out opportunities for informal hallway discussions
- Staying apprised of education research relevant to the discipline, and conveying this to the faculty
- Helping set up training programs for TAs within the department to allow TAs to better support the use of new teaching methods

Position Description

At both institutions there was an initial difficulty with the formal job title and description for the SES. Existing university job titles allowed for two explicitly distinct types of short-term PhD-level positions: research postdocs and instructors. The formal policies of both universities forbade a non-tenure-track person from doing both teaching and research (including research on teaching and learning). A person who helped with teaching a course, did research on the effectiveness of the teaching in a course, and might choose to publish that work (as SESs were expected to do) was in conflict with some aspect of every official position description. I had to negotiate a change in official position descriptions at both universities in order to make it possible to hire SESs.

SES Candidates

There was little difficulty in finding excellent candidates for SES openings except for in computer science. In the standard model, good candidates were new PhDs in the respective disciplines with people skills and a strong interest in education. In computer science (CS), the model included buying out some of the teaching load of suitable CS instructor-track faculty so that they could serve in the SES role. This worked well.

Selection and Hiring

It was important that the departments have ownership of the SES recruitment and hiring process, although SEI Central always participated in the hiring process and interviews in an advisory capacity. There were variations across the departments as to how hiring was carried out. Usually there was some form of open search for an external candidate, but in some cases the department had internal candidates (recent PhD graduates or sessional instructors) they felt would be well suited. I have no reason to argue for one over the other, as I have seen that both can work out well. Because it took time for an SES to become effective, and it takes time to transform multiple courses, at least a two-year appointment was considered essential and a three-year appointment was preferable.

In departments where educational activities had particularly high public visibility (for example, because of substantial educational research in the discipline), the applicant pool for SES positions was quite large (forty to fifty applicants, with approximately half those being worth serious consideration); in other departments, the applicant pool was smaller (ten to twelve). We were pleased with the quality of the top candidates in nearly all cases. In the later years of the SEI, it was increasingly possible to hire postdocs in some disciplines who had both a background in the relevant discipline and science education expertise. Such a background is not sufficient to ensure that an SES is effective, however, as many other skills are also required.

Advertising was done through a wide variety of channels, including disciplinary research, education research, and teaching- and learning-related venues. Disciplinary-specific search channels (for example, advertisements in a professional society journal) did not typically attract many suitable candidates. Advertisements described the position and its duties, which included working with faculty to develop course materials and measures of student learning. A PhD was typically required, as were organizational, interpersonal, and communication skills, with experience in education listed as a plus. Most positions were advertised as one-year renewable appointments. Examples of advertisements are available in Appendix 3.

Departments typically invited the top candidates to visit the campus and give a talk on their research and/or an education-related topic. Interviews often included questions about their interest in the position and relevant expertise (that is, disciplinary knowledge, education, and education research). The most important criteria were whether the applicant's personality and work characteristics were a good fit for the position. For example, candidates were often asked how they might handle a scenario, such as a faculty member who is resistant to change. Red flags would include a candidate who suggested that the faculty member just needed to be convinced of the effectiveness of the change, or who expressed overconfidence about his or her knowledge of science education, rather than an interest in and willingness to collaborate, listen, and learn. Other questions might ask candidates to describe a time when they did not feel adequately supervised or had to deal with a difficult person.

At SEI Central, we participated in the hiring process in several ways. First, when the person who was handling the search did not have experience in the process, we had preliminary discussions with that person about factors that were important in selecting a good SES, and we helped write the advertisement and a brief job description. In some cases we offered suggestions about places to advertise for candidates. We were always involved with the interviews, meeting with the candidates and providing our suggestions to the department, but always deferred to the department when it came to making the final decision. (There was never a case of a serious disagreement over the choice.) An early lesson learned was that during the interviews with candidates, we needed to discuss quite explicitly the relationship between SEI Central and the department. This included stressing that the SES would work for the department and have a primary supervisor in the department, but that SEI Central provided the salary money and had a small set of requirements the SES needed to follow: participation in training, reading group, SES meetings every one to two weeks, and submission of progress reports (originally every two weeks, later once a month). We also discussed the resources and assistance we would provide to them.

SES Course Transformation Activities

In this section, I list the major components of the SES role, including lessons learned as to how SESs could be most productive within that role. Since every situation was somewhat different, the relative emphasis of these components differed, and no single SES was likely to be heavily involved in all of the listed activities. SES activities were organized around three separate guiding questions of the SEI model of course transformation: What should students learn? What are students learning? Which instructional practices will improve student learning?

What Should Students Learn?

In order to answer this question, SESs undertake several activities, as described below.

Develop learning goals. Learning goals define operationally what students should be able to *do* as a result of learning about the content.[1] Ideally, the SES would meet with individual faculty members to find out what their overall learning goals were for the class. What were the big ideas that the faculty were looking to get across to students? What did they feel that the course was "about"? Did they have goals that were not content-specific, such as developing critical thinking or improving student interest in the topic? What knowledge and skills were needed for follow-up courses?

Useful approaches taken by the SESs were:

- Asking the instructors for examples of student work that demonstrated to them where students were and were not achieving the desired understanding
- Going over past exam questions with the instructors and asking them to explain why they included the question and what they felt it was testing
- Asking instructors of subsequent courses in a sequence what they noticed that students could not do that they wished or expected they would be able to (surprisingly, instructors of subsequent courses were often better able to articulate learning goals for the preceding course than was the instructor of that preceding course)
- Providing relevant examples of learning goals
- Working with faculty in facilitated groups to develop learning goals

Below are several SES prompts for use in discussions with faculty that worked well to elicit faculty ideas about their instructional goals and needs:

- After you lecture on this topic, what do you expect a student to be able to do?
- If a student gets this exam question right, then what does it show that the student can do?
- What do the students have the most difficulty with? What would the students do that would show you they got it?
- What are things that students have said or done that indicated to you they did not get it?

As discussed in Chapter 6, developing learning goals is not necessarily the best way to start working with faculty, because it is difficult and does not provide immediate rewards. It is, however, an important step in a course transformation. Learning goals are valuable because they allow the faculty to more effectively target instruction and assessment, and they enable communication with both students and fellow faculty about course expectations. Learning goals were almost always modified and improved after the first iteration of the transformed course.

Facilitate consensus among faculty. The original vision of the SEI was to create faculty working groups, which would collaborate to develop learning goals and review assessments. Such dedicated working groups functioned well in only a few departments, such as physics at CU, where such discussions were part of a preexisting culture of teaching and learning in the department.[2] Even when establishing such a working group is not successful, it can be productive for the SES to gather some relevant faculty for one or two meetings to discuss outcomes of course transformation. Typically, several faculty members were interested because they would be teaching the course concurrently or in a later semester, had taught the course in the past, or were teaching a follow-on course.

In order to lead such a meeting of faculty, the SES needed to have good facilitation skills, including the ability to actively listen. The book *Getting to Yes* by Roger Fisher and William Ury is a useful guide for working with faculty members, and it is one of the books that SESs were given when they first started the job. Additionally, several things that did and did not work well for SESs in facilitating faculty meetings are outlined in Table 4.1.

What Are Students Learning?

SESs engaged in a variety of activities to generate data to drive course transformations. These activities were typically undertaken in collaboration with the faculty member(s) teaching the course, or the faculty working group, if one existed.

Identify students' prior knowledge. What knowledge and skills should students have (or what knowledge and skills are they assumed to have) at the beginning of the course? SES methods for identifying and assessing such prior knowledge included conducting interviews with faculty members; searching the discipline's education research literature to identify relevant student ideas or misconceptions; and developing diagnostic pretests, homework, or other activities to measure student knowledge upon entry into the course.

Identify student learning difficulties. Where do students tend to struggle with the content? These are the key areas where course development should focus. SES methods included four ways to identify learning difficulties. First, interviews conducted with faculty who had taught the course or subsequent courses, to determine which topics or skills students had the most difficulty with. Second, searching the discipline's education research

Table 4.1. Do's and don'ts for meetings with faculty members

To productively lead a faculty meeting . . .	
Do . . .	Don't . . .
Meet with faculty individually to identify their personal priorities and concerns	Treat the group as the only source of input, or as a singular unit
Encourage broad participation, inviting the entire faculty and targeting individual faculty members	Rely on mass emails alone
Distribute a clear agenda and other materials in advance	Be too rigid in following the agenda
Choose a topic that will motivate faculty to attend	Call a general meeting without a topic of broad or urgent interest
Designate a knowledgeable facilitator who can guide and synthesize discussion	Hold a meeting with no leader/facilitator, or have a leader who is focused on expressing his or her own opinion
Approach discussions in the spirit of soliciting faculty guidance and input	Proselytize about education
Discuss course objectives and pedagogical issues	Create the impression you are telling faculty how to teach
Send out summaries of meeting accomplishments	Assume faculty will remember or recognize the progress made
Hold several meetings	Rely on a single meeting
Synthesize meeting results and produce working documents for circulation and discussion in the next meeting	Expect most faculty to consistently do homework
Survey faculty to establish areas of consensus and priority (for example, rate the relative importance of learning goals). Ensure they have an opportunity to express views, even if they choose not to	Expect to reach clear consensus through discussion
Follow up with faculty about how their input has been used	Move ahead with the project without letting faculty know the outcomes of their investment of time

Source: Adapted from Rachel E. Pepper, Stephanie V. Chasteen, Steven J. Pollock, Katherine K. Perkins, "Facilitating Faculty Conversations: Development of Consensus Learning Goals," *2011 Physics Education Research Conference* (Melville, NY: American Institute of Physics, 2012), 291.

literature for studies on student learning in the topical area. Third, examining existing course data (for example, homework, tests, surveys) for insight. Finally, it was always revealing to observe, survey, and interview students.

The last item, collecting data from students, was an important part of what many SESs did. This included observing students during class, help hours, and/or discussions, particularly noting student questions. Another data source was student attitude surveys, including asking what they found most useful about the course, or how they viewed the course and its content.[3] Finally, the most in-depth examination involved individual or group interviews. Student interviews were typically done in a cognitive "think aloud" format as the students worked through problems or questions. (See Appendix 2 for a guide to interviewing both students and faculty that was used in the SES training.) Conceptual assessments were carried out by administering validated instruments and short formative assessments (such as two-minute papers or short, targeted questions created by the SES) during class or at the end of the course.

Many of these activities and questions naturally led to research on student learning. Before the SES embarked on such a research study, it was helpful to make sure that the data would be of interest and use to the faculty members. Thus, the SES began by asking faculty members whether there were any data on student learning or attitudes that they were particularly interested in seeing.

Develop measures of student learning. An important part of the SES job was assessment, obtaining measures of student learning to determine the effectiveness of the transformation of course materials and teaching. This assessment data could take many forms. One form was student scores on traditional assessments (for example, exams and homework), although care had to be taken in using typical faculty-prepared questions as they were often not very meaningful. Successfully solving them often involved knowing some obscure trick, or they could be solved by simple memorization of facts or procedures without much grasp of the material. Other forms of assessment include student responses on feedback and attitude surveys (both midterm and end-of-term); student scores on conceptual assessments, validated or not; observation of the course, either through field notes or by means of validated observational protocols, such as student engagement or instructor practice;[4] and other course data, such as drop/fail/withdraw rates, attendance, or persistence in major.

Ideally, such data would be acquired both before and after the transformation of each course. However, as discussed in Chapters 3 and 6, it was difficult to obtain baseline data on student learning (that is, measures of student learning prior to course transformations), which would allow comparison to post-transformation results.

We initially encouraged the SEI departments to develop and validate instructor-independent measures of learning, as described by Adams and Wieman.[5] Over time, however, we reduced our emphasis on such conceptual assessments, because the level of effort and expertise required to develop them was too high relative to the value placed on such assessment data by the faculty. The one case in which such assessments were developed and routinely used as envisioned by the SEI was in the CU physics department, in which SES time was devoted to a single course over multiple years, and such work was supported both through the existing physics education research group and external grants.[6] A few other tests of conceptual mastery and attitudes about learning were developed as part of SEI activities, but the degree to which they were used is unclear. In most other departments, if the instructor had developed some reasonable measures of learning (usually in consultation with the SES) that could be used repeatedly, this worked fairly well, even without independent validation of the assessment.

What Instructional Approaches Improve Student Learning?

The next step of the SES job was to decide on the methods and materials that would be used to better teach the content. During this phase, the SES collaborated closely with faculty, faculty teams, and TAs. Typically, the SES played a larger role in material development at the beginning of the course transformation process, gradually transitioning into a more advisory role as the project progressed, providing feedback on materials developed by the faculty.

Develop curricular materials and teaching approaches. The SES began by finding out what the faculty were most interested in—what were the educational challenges they wanted to overcome, and what teaching skills did they want to develop? The SES acted as a knowledgeable coach during this phase of the project; it was important to avoid coming across to faculty as "preachy," or as having an agenda. Letting faculty interests drive the collaboration was one way to achieve a productive partnership.

Next, the SES might describe a variety of teaching approaches that could be used (such as clicker questions with peer instruction, in-class worksheets, or case study teaching), and give the faculty member an opportunity to observe these methods in action in another course.

Using the learning goals as a guide, the SES could then develop a variety of materials for use in specific classes (for example, clicker questions, worksheets, tutorials, invention activities, case studies) or out of class (for example, homework, recitation activities, tutorials, labs). This was always done in collaboration with the faculty member, who made the final decision as to what would be used.

Lastly, the SES could provide the instructor with feedback on short and long-term student outcomes based on their scores on assessments and on classroom observations (see below). I discovered that with any research-based teaching method, there are countless possible ways to implement it badly. This was particularly likely to happen when the instructor did not understand the underlying principles of learning on which the method was based. A large part of the SES's job was to master these principles and guide the faculty member in how to avoid the pitfalls.

As part of their training, SESs learned about many common mistakes and how to avoid them when implementing new teaching methods, and passed this guidance on to the instructors. (This list of common mistakes and good practices grew substantially over the course of the SEI, based on SES observations. See Appendix 1.) This SES support in avoiding many early unpleasant stumbles as the faculty members adopted new teaching methods played a large part in the success of the SEI.

When applicable, the SES could help co-teach some of these activities in class—giving the SES more direct experience with student interaction in the activity, and providing the faculty member with additional instructional support and an opportunity to observe unfamiliar teaching methods in operation. While such co-teaching in the process of implementing new materials and methods was encouraged, it was necessary for SEI Central to have oversight and to define restrictions (discussed below) to ensure that SESs were not simply used as replacement instructors.

Observe the transformed course. The SES typically observed the classes in the transformed course and provided ongoing feedback to the instructor based on those observations. Again, it was important for the SES to develop a supportive, coaching relationship with the faculty member, so that they could function as a partnership. To achieve this, the SES could focus on giving feedback that (1) related directly to areas where the faculty member had already expressed interest, (2) had the greatest potential for improving student learning, or (3) had the greatest potential for changing the faculty member's perspective on teaching (for example, suggesting ways to get students more intellectually engaged with a concept). This was arguably one of the more sensitive aspects of the SES job, and they received substantial training and support in developing positive approaches to faculty coaching, especially as our own understanding of these best practices evolved over time. It was important for the feedback/discussion with the instructor to occur very soon after the observed class. Brief feedback immediately after class, when the class was fresh in the instructor's mind, was more useful than a detailed meeting a few days later.

SESs typically found it difficult at first to know what to look for in class observations out of the vast assortment of things they could be watching, and so this was an important part of their training. Although the most useful feedback tended to involve specific details about how particular issues or student questions and concerns were handled in a given class, and what they could learn from watching and listening to nearby students, it was useful to develop some standardized observation protocols.[7] These allowed the SESs to quantitatively and reliably characterize student engagement and how the faculty and students were spending their time during a class period. Such quantitative numbers could sometimes be more effective in convincing faculty to change their practice than subjective feedback from the SESs, particularly if the quantitative results were surprising to faculty. For example, the Classroom Observation Protocol for Undergraduate STEM (COPUS) observations showed some faculty that although they intended to use active learning methods, they spent more class time lecturing, with students passively listening, than they had realized.

Archiving and Disseminating the Results

In order for course materials to be used by others, they must be archived and shared with the community—both within and outside of the department. Thus, part of the SES role was to create a course materials package that would be available for use by instructors in the department and in the broader education community. Typically, this archiving task was undertaken after the second iteration of the transformed course.

Faculty indicated that they wanted to have materials arranged so that they could easily pick and choose what they wanted to use, rather than have to search through an entire package. Significant time and effort were devoted to creating an online course materials management system so that materials across departments and institutions would be centralized and organized into a common structure.[8] It was challenging to create a model that worked for all possible cases and was easily used; regrettably, this online structure served more as a resource for SEI staff than for faculty. Among instructors, course binders (either as electronic zip files or as paper binders) were still the mainstay. SESs were extremely helpful in creating this organized archive, because existing departmental structures and expectations provided no incentive for a faculty member to expend the necessary effort to document, organize, and communicate the course changes for an external audience, including other faculty in their department. Once the SEI funding ended, however, there was no clear mechanism or responsibility for maintaining these archives in a department.

Another aspect of dissemination involved presentation and publication of research papers on course assessments, research findings, student learning, course transformation, or other aspects of the SEI work. SESs and departments were told that an SES was expected to publish at approximately half the rate of a regular research postdoctoral fellow in the department. This expectation was set for two reasons: (1) professional development and status of the SES, and (2) establishing a standard for the quality of work done by the SEI as suitable for publication in a peer-reviewed science education journal.

It was a continual challenge to get SESs and departments to meet this publication expectation, primarily because it was not well aligned with either of their priorities. The publication of educational research was not seen to be of great importance. I would regularly encourage SESs to do so, but with at best limited success, except for the few who saw such publication as important for their future careers (those who planned to become faculty members doing education research). That said, the current total of more than 120 publications (www.cwsei.ubc.ca/SEI_research) across both institutions is significant and has contributed substantially to the literature on educational change and student learning within and across disciplines.

In addition, there is a substantial amount of unpublished work generated by the SEI that could also be a contribution to discipline-based education research—but will likely never be published.

Local dissemination of SEI results was clearly beneficial to the SEI efforts within departments and was practiced regularly by the SESs (often in collaboration with faculty members) in the most successful departments. This dissemination took several forms: monthly newsletters describing SEI activities and notable results, verbal reports at faculty meetings, more extensive write-ups provided in advance of discussions at departmental retreats, and departmental colloquia and seminars on notable SEI work. The last of these were usually presented jointly by an SES and a faculty member. As well as distributing the newsletters in the usual manner, it was found to be productive to prominently post them where they would stimulate discussion, such as in the faculty coffee room and right outside the door to the departmental office.

SEI Central also ran an annual end-of-year mini-conference at which all the SESs and some faculty would present posters on their work. All of the SESs were very involved in this event, usually presenting multiple posters. This event would bring in a limited number of faculty who were not involved in SEI work to learn more about the SEI activities and results in their own department. These events were particularly successful at bringing together faculty and SESs already involved in SEI work from across the departments for discussion. An added benefit of this conference was that the posters were then uploaded to the SEI website, providing a public archive of the SES work (see, for example, www.cwsei.ubc.ca/EOYevent2014.html).

In a few cases, dissemination also included creating written and video-based materials aimed at helping faculty use a variety of instructional techniques. For example, videos, workshops, and a booklet were developed by SESs for helping faculty use clickers and learning goals effectively, and all have been cited and used beyond CU and UBC.[9]

SES Responsibilities

The SES position is unlike any that traditionally exists within a department, and most existing positions—such as instructor, course support personnel, or researcher—provided a misleading model for the vision of the SES as an embedded expert in education. Over time, SEI Central found that clearly defining the SES role made for a more productive experience for all involved.

One lesson learned (which resulted in program improvements at UBC compared to CU) was to make the SES role and responsibilities clearer to departments and to the SESs themselves at the point of initial hiring.

SES training and meeting attendance. We needed to clearly indicate that attendance at the weekly SES training and reading group meetings was mandatory. It was not realistic to assume that SESs would be able to quickly develop the necessary skills for such a complex job simply by reading relevant articles and books. At UBC, both the meeting expectations and the training program was much more formalized than at CU, with regular schedules and expected deliverables for training exercises. These expectations were mentioned during the job interview, explained to new SESs when they arrived, and communicated clearly to the departmental director. These clear expectations were important for ensuring that the necessary training was given priority, especially in light of all the other time demands that the SESs encountered. As discussed below, this structure also contributed to a more cohesive, supportive SES cohort.

Balancing work demands. One of the most demanding aspects of the SES position, and one that all new SESs struggled with initially, involved balancing the demands of training and learning, working with multiple faculty members, and producing material for courses in a timely manner. In the early days, SESs experienced a great deal of frustration around juggling these multiple demands, in part because the expectations had not been made sufficiently clear to them and to departments. They did not always know what they needed to do in order to do a "good job." While the job always required good time management skills, the frustration associated with the multiple demands largely dissipated over time—likely due to the various adaptations that were made to the program, such as improved departmental planning structures and SES supervision and training. In particular, both the departments and the new SES were advised that their first semester should be spent on a small project rather than a full course transformation, as the latter was overwhelming at that stage. An additional factor in improving SES job satisfaction was the presence of the SES community (about fifteen to twenty SESs during the most active years at UBC) that communicated expectations and other guidance to new SESs.

SES activity reports. SESs were required to provide brief (one-page) reports of their progress (initially every two weeks, later once per month). These reports went to both SEI Central and the SEI department director. These were reviewed by SEI Central with particular attention to:

- Whether the SESs were planning properly and dealing effectively with the large number of different demands on their time, or becoming overwhelmed
- Whether the department was paying attention to what was being expected of the SESs, or whether multiple faculty members were putting demands on them with no central coordination or oversight
- Whether any of the SESs were working on something for which there was research literature, prior SEI work, or people who could be helpful to them but which they didn't know about
- Whether they were spending time productively and not wasting time due to poor work habits or lack of cooperation or support from faculty or the department

The fact that all of the issues listed were encountered fairly regularly made it clear that such reports and responses to them were needed. Having such reports in hand during the meetings between SEI Central and the departments also made those meetings more focused and productive.

SES Supervision

An ongoing challenge was establishing to all concerned how the SESs fit within a chain of command—to whom they would report, and who would be responsible for managing their priorities. It was vital that SESs be seen as members of the department and resources to its faculty. In cases where faculty members viewed SESs as thrust into the department by myself or the university administration to "fix" departmental instruction, the results were predictably poor.

However, it was also important that SEI Central be able to provide oversight to ensure that SES time was being used effectively, that departments were providing adequate supervision, and that the SESs received the necessary training and professional development to be successful. In some cases, SESs became so engrossed in their daily activities and the demands of course transformation that they neglected the training and meeting requirements—which had a negative impact on their performance. In other cases, departments sent them off to work with unwilling faculty members with no help or guidance.

Thus there was a continual tension between SEI Central and the departments in terms of who controlled the SESs' time. Laying out clear, explicit expectations, as described above, along with providing a formal training program and clear supervisory structure within the departments, was very helpful in this regard, but the issue required constant attention. It was important to be helpful and supportive of the SESs while being quite explicit, to both SESs and department directors, as to what issues and decisions were the responsibility of department directors and not SEI Central.

The SES and departmental activities were monitored through the SES meetings, email reports, other communications, and the regular meetings between SEI Central and each department (including the departmental director and SESs). In order to keep responsibilities and lines of authority clear, we had meetings with the department's SEI director and the SESs, and occasionally department chairs, to explicitly go over which issues SEI Central would *not* provide input or decisions on, and why these should be handled by the department. On rare occasions, this would also mean discussing with an SES and a department director what was expected of the SES with regard to SEI Central activities—for example, that the SES was required to attend important training sessions and provide required reports.

SES Morale

In the early days, many SESs arrived excited to have been hired to make improvements in teaching in the department but soon became very frustrated. As described elsewhere, it was not uncommon for a department to assign an SES to transform a course but overlook the fact that the faculty member teaching that course was not interested in working with the SES and/or changing their teaching methods. I expended a lot of time and effort trying to preserve SES morale under such conditions. Even with this effort, approximately 25 percent of the early SESs quit before the end of their appointment, usually after about one year. About 25 percent of the rest of the early SESs were on the verge of quitting. As the expectations for department management of SEI activities became more clearly established (for example, the department set expectations and made agreements with a faculty member *before* sending the SES to work with that person), the SES training program better addressed specific issues (such as faculty resistance, common SES experiences, and appropriate expectations), and a more vibrant SES community grew over time, this attrition became much less of a problem. In the later years of the SEI, nearly all SESs remained for two or more years, often leaving only after being recruited for very attractive long-term positions, with our blessings.

SES Teaching Responsibilities

Initially, a rule was established that SEI-supported SESs could not have primary responsibility for teaching a course—that is, they could not be the instructor of record. This was done to prevent them from simply being used as free replacement instructors. Having a highly qualified instructor who is not paid from department funds is very tempting, especially to a chair who is grappling with budget problems and not particularly supportive of the SEI work.

This restriction was modified when it became apparent that teaching experience was an important part of SESs' professional development, both to help them to do their current job well (increasing their credibility in the department and giving them valuable experience to draw upon) and as résumé experience for future positions. Allowing SESs to teach had the added benefit of providing a model for faculty of how they might use various teaching techniques—SESs regularly invited faculty to observe their own classes.

Thus SESs were allowed to, and regularly did, teach as the instructor of record for courses, with the proviso that SEI funds would not be used to pay their SES salary for that time, and that an SEI-supported SES would not teach more than one course a year. It was necessary for SEI Central to monitor such situations fairly closely to avoid exploitation of the SES and misuse of SEI funds.

How to Work Effectively with Faculty

As described before, it was important that SESs act as partners and gentle coaches to faculty. Below are several elements of effective faculty partnership that worked well.

Developing and communicating scholarly expertise around teaching and learning. SESs who took their roles as educational scholars and departmental resources seriously were particularly effective. Faculty came to recognize that those SESs had valuable and unique expertise, and this resulted in more effective working relationships. Many SESs have commented on the importance of having both disciplinary and pedagogical expertise in achieving the respect of the faculty and establishing good working relationships. Additionally, in the few cases where an SES's disciplinary expertise was weak, that individual's effectiveness was noticeably reduced.

One example of dissemination of scholarly expertise was the monthly newsletter produced by the Department of Earth, Ocean, and Atmospheric Sciences, the *EOAS-SEI Times*.[10] Designed to be easily skimmed, these two-page documents with titles such as "An Instructor's Clicker Cheat Sheet," "Making the Most out of the First Day of Class," and "Department Feedback about EOAS-SEI" helped to inform faculty about relevant literature and best practices, as well as SEI efforts in the department.

Finding interested faculty. Originally it was assumed that at the proposal stage departments would identify a list of courses to be transformed, and that this would serve as guidance for SES work. In some cases, with strong and consistent departmental leadership, this model was followed productively (see Chapter 5 for a noteworthy example from EOAS).[11] The SESs then systematically worked through a list of predetermined courses to transform.

In many cases, however, SESs discovered that faculty teaching those courses previously identified for transformation were not necessarily interested in the course transformation efforts. In such circumstances, it proved more productive to have the SESs work with individual faculty members who were interested in making changes in their teaching rather than working on a particular course. This represented a shift from *course-focused* work to *faculty-focused* work. Initially we were rather nervous about this, as it implied an abandonment of the model of departmental ownership of courses. We became more comfortable with this approach as we saw that the more faculty members who were engaged in thinking about and changing their teaching, the more the culture of the department with regard to teaching was changing. This, in turn, resulted in more faculty spontaneously deciding to learn about and adopt new teaching methods.

SES Training and Community

Originally the naïve assumption was that new SESs would be able to develop adequate skills by reading books and articles, applying those ideas in practice, reflecting on the experience, and engaging in further reading and discussion in informal meetings. This was not generally the case. SES training was formalized over time by providing a more consistent and coherent training experience. While numerous models were tested, in this section I present the version of SES training that was found to work best. It included a new SES development series, reading group, and ongoing regular SES meetings.

One difference between UBC and CU that impacted the training was the *number* of SESs in each program. At CU, there were fewer total SESs, and few new SESs were hired after the initial program initiation. At UBC, on the other hand, the program was larger, and there was a new cohort of SESs each year. This made it more feasible to support regular initial and ongoing training for SESs, which created a greater sense of community and collaboration among SESs at UBC. It also made it possible to involve experienced SESs in the training of incoming SESs, which had multiple benefits. Thus, the impact of SES training, the resulting SES capacity, and the sense of SES community was significantly greater at UBC than at CU.

New SES Development Series

During the first semester after being hired, SESs engaged in a structured, one-semester seminar and discussion series, the STLF Development Series (STLF being the UBC name for SES). The series consisted of approximately one 90-minute meeting per week for twelve weeks. Each week, SESs would read an article or section of a book in advance of the meeting. The primary texts used were *How People Learn*, by John Bransford et al., *How Learning Works*, by Susan Ambrose et al., and, as mentioned above, *Getting to Yes*, by Roger Fisher and William Ury.[12] During the meeting, they would discuss the reading and work in small groups to put the lessons into practice, such as applying the strategies to create activities for courses with which they were involved and discussing the proposed activities.[13] An abbreviated list of topics covered included:

Effect of prior knowledge
Knowledge organization: expert vs. novice
Motivation
Learning and transfer
Deliberate practice

Development of self-directed learners
Learning goals
Formative assessment
Memory and retention
Peer instruction and effective clicker use
Group work
Characteristics of expert tutors

These weekly assignments and discussions were directly linked to things that would be done in a course transformation, and included analyzing the principles and research behind the activity design, as well as SEI Central staff providing feedback on their work. There was also considerable discussion and guidance in the training about how to work most effectively with faculty. The first semester of an SES's work was typically devoted to planning a course transformation and to the Development Series.

The schedule of the training program for new SESs was made available to existing SESs, which resulted in many coming to specific sessions. By the end of the SEI program, many of the weekly training sessions were facilitated by senior SESs, which greatly enhanced both SES community and capacity. In addition to lessons in teaching and learning, senior SESs were able to help their newer colleagues navigate the often subtle aspects of the job and set realistic expectations. I learned that the few SESs who had significant training in education research (including PhD-level training) still needed to go through the training program to be effective, although this was not always obvious to those SESs themselves.

There were often non-SES people who were interested in the SES training program (such as new faculty or instructors), and participation was encouraged. As a one-time experiment, we tried having an abbreviated SES program specifically for new UBC science faculty, but it was not very successful. Although new faculty members did sign up (in response to encouragement from the dean), attendance and completion of assignments were quite erratic.

SES Meetings

In addition to the new SES Development Series, SEI Central held a meeting with all the SESs every two weeks. These meetings provided ongoing professional development for SESs, facilitated the sharing of ideas and resources, and built community. They also provided a venue where SESs could easily discuss and seek help from SEI Central on sensitive issues within their department (usually involving difficult interactions with faculty). Meeting topics varied: discussion of new research studies in the literature, designing effective research studies, data analysis, designing effective instructional activities, conducting cognitive interviews with students, sharing experiences of what worked well in a department (or not), and presentation of work by some of the SESs, particularly when they had tried some novel method and had data on the results.

Science Education Reading Group

SEI Central also ran a reading group that met every two weeks, in which a science education or cognitive psychology paper was discussed. The reading group included a number of faculty members and graduate students as well as the SESs. The focus of the papers varied, with topics including valuable teaching methods that had good supporting data, fundamental research about learning and brain science, and examples of good and bad research papers to help SESs and faculty in thinking about carrying out and publishing research on their own educational activities. A particularly valuable aspect of the reading group was the online Basecamp tool that led to the large virtual reading group, discussed below.

In-Person Community Building

The SESs at UBC developed into a cohort, working together and helping each other, both within and across departments. There were many factors that contributed to this happening much more at UBC than at CU. This included larger numbers of SESs, regular meetings, regular readings and frequent use of discussion group, regular social events, a good meeting space more connected with SEI Central, more management oversight, having existing SESs participate in training of new ones, and to some extent the personalities involved, as some individuals took it upon themselves to develop a community.

Online Community Building Tool

SEI Central also provided various activities aimed at building community among the SESs. One helpful tool for this was the use of a commercial project management tool called Basecamp. Basecamp allows for threaded discussion, file attachments, and email notifications to users, among other functions. Basecamp thus provided a central location for SESs (across both institutions, to some degree) to ask questions, discuss specific topics and to share papers and other materials, and for SEI Central to quickly send guidance and resources to all SESs at once. As new SESs joined, they were added to Basecamp, and then could use the previous conversations and posted materials as a resource. We frequently were able to answer questions from new SESs just by referring them to the existing materials on Basecamp.

At UBC there was ongoing involvement on Basecamp of UBC SEI alumni (that is, former SESs) who had gone on to other jobs and institutions. Former SESs typically remained on Basecamp, and they would continue to contribute (at a reduced rate) to discussions, providing advice and materials, and letting current SESs know about job openings. As SEI alumni grow in number and have spread throughout Canada, the United States, and beyond, this online community provides a valuable resource for current and former SESs.

One portion of Basecamp that has been particularly valuable is the virtual reading group. This group was originally started as a way to provide materials for the in-person reading group to prepare for each meeting and facilitate ongoing discussions and sharing of related materials. An increasing number of people have signed up for this group, including CU and UBC faculty and SESs, and this virtual group now numbers over two hundred. Although only a small fraction participate in discussions of papers, we find that a much larger number regularly read the papers and comments. Basecamp also allows new users to easily access previous papers and the discussions around them.

Career Paths of SESs

Initially it was difficult to know what the long-term career path for SESs would be, and there was some concern as to whether the experience was a suitable step toward a successful long-term career. However, experience has shown that SESs have had desirable career options, and in many cases are able to choose among multiple attractive offers. There are clearly viable career paths for individuals with this training and experience.

In Canada, where tenure-track teaching faculty positions are fairly common, that has been the most common career path for UBC SESs. They have proven to be *very* competitive for such jobs, as they bring a unique level of expertise in learning and teaching in their disciplines.

In the United States, there has been somewhat more variation. A few SESs have gone to college or university tenure-track positions with a focus on teaching, and others to long-term untenured teaching positions—sometimes in the department for which they worked as an SES. A few SESs have gone into tenure-track faculty positions in science departments, carrying out research in discipline-based education. A number have also been hired to run university centers for teaching and learning, or the science portion of such centers.

FIVE

What Was Achieved and What We Learned

THE SEI WAS fundamentally an experiment in institutional change and much was learned from that experiment. This chapter presents the rich set of data that was produced and all that the experience revealed. This ranges from detailed measurements of changes in teaching methods used in various courses to subtle observations of how departments function and oversee their courses, and how these differences impact the quality of teaching. Included in the results are discoveries of how institutional structures and values make it very difficult to collect some types of data, particularly instructor-independent measures of learning. In this chapter I show the substantial changes in teaching that were achieved, the contrasts in achievement across departments, and the differences that were responsible for these contrasts. This reveals a number of traditions that had inadvertently become established in individual departments and that negatively impact educational quality. The information in this chapter reveals the many opportunities for teaching improvement and the myriad issues that can interfere with educational improvement at this level, as well as many clever approaches developed by departments to make their SEI efforts successful.

As discussed in Chapter 3, the implementation of the SEI revealed unexpected challenges in collecting data. However, many types of data were collected on the impact of the SEI and how this varied across departments and institutions. This data include some results on student learning outcomes and evaluations, and extensive data on the adoption of new teaching methods, including the number of courses and faculty using these methods. There are also many observations about the relevant respective departmental organizations and functioning and how these impacted the success of SEI activities. I also present information on the changes in the departmental cultures at the different institutions. Lastly, I present some information on economic issues, including the impact of the SEI on instructional costs.

Student Learning Outcomes

In spite of the data collection difficulties, there were many examples where student learning outcomes were measured for individual courses and for particular learning activities, often with comparisons with outcomes from previous iterations of the respective course. Collection of these data were usually instigated and carried out by SESs, particularly those interested in pursuing careers in science education research. Many of these have been published or presented at conferences and are in the list of 120+publications of the SEI at www.cwsei.ubc.ca/SEI_research. Some representative examples are listed in Table 5.1.

There are also a number of additional examples that have not been published. In nearly every case these examples showed that when research-based instruction was put in place in the SEI, it resulted in improved student learning. The few exceptions usually involved courses where there were very serious problems with the basic content and organization (see "Curriculum Issues" in Chapter 6). Those examples demonstrated that if a course is badly designed, the quality of pedagogy makes little difference. Generally, such a course was a large, apparently random, collection of topics joined together for ancient and unknown reasons and misaligned with student preparation.

Student Course Evaluation Results

The institutions collect student course evaluations in every course. These data were found to be of little value, both because of the general limitations of student evaluations and the fact that the questions on the student evaluation forms used at both UBC and CU were of questionable design.[1] As a result, the primary interest in looking at the student evaluations was to check if the transformed courses had lower or higher evaluations than their traditional counterparts. It is frequently claimed, though usually with little supporting data, that

Table 5.1. Published examples of SEI measurements of student outcomes

Title	Reference
Why peer discussion improves student performance on in-class concept questions	Smith et al., *Science* 323, no. 5910 (2009): 122–124
Using invention to change how students tackle problems	Taylor et al., *CBE—Life Sciences Education* 9, no. 4 (2010): 504–512
Learning and retention of quantum concepts with different teaching methods	Deslauriers and Wieman, *Physical Review Special Topics: Physical Education Research* 7 (2011): 010101
Improved learning in a large enrollment physics class	Deslauriers et al., *Science* 332, no. 6031 (2011): 862–864
The Colorado Learning Attitudes about Science Survey (CLASS) for use in biology	Semsar et al., *CBE—Life Sciences Education* 10, no. 3 (2011): 268–278
Successful curriculum development and evaluation of group work in an introductory mineralogy laboratory	Dohaney et al., *Journal of Geoscience Education* 60, no. 1 (2012): 21–33
Teaching methods comparison in a large calculus class	Code et al., *ZDM Mathematics Education* 46, no. 4 (2014): 589–601
Educational transformation in upper-division physics: the SEI model, outcomes, and lessons learned	Chasteen et al., *Physical Review Special Topics: Physical Education Research* 11 (2015): 020110
Teaching critical thinking	Holmes et al., *Proceedings of the National Academy of Sciences* 112, no. 36 (2015): 11199–11204
Transforming a fourth-year modern optics course using a deliberate practice framework	Jones et al., *Physical Review Special Topics: Physical Education Research* 11 (2015): 020108
Teaching students how to check their work while solving problems in genetics	McDonnell and Mullally, *Journal of College Science Teaching* 46, no. 1 (2016): 68–75

introducing active learning methods into a class will result in student course evaluations going down. This was a fear often heard from faculty.

Sampling of student evaluations for faculty at both institutions showed that a faculty member's student evaluations typically remained unchanged (within statistical uncertainties) from before to after SEI course transformations. This was true even though in most of those cases the teaching methods were dramatically changed and in many cases data showed substantial improvements in learning. As noted in Appendix 1, faculty received specific guidance on how to get student buy-in for research-based teaching methods. Without this guidance, the student evaluation results might have been different.

There were a few cases where there was a notable decrease in the evaluation score. To my knowledge, these all involved cases where a faculty member made a large number of changes in a course all at once without, in my judgment, adequate preparation. Students rated the instructor significantly lower than in previous years and commented that the course was disorganized and poorly prepared (a sentiment shared by the associated SESs). However, the evaluations then rebounded in the following year, when the courses were presumably better prepared.

In the one department (UBC EOAS) where there was the most widespread shift in teaching methods, the teaching evaluations across the entire department were compared after roughly half the courses were being taught in transformed fashion. From that data, it appeared that student evaluations of the faculty who had altered their teaching remained unchanged from the pre-SEI period, but the evaluations of the faculty who had not changed their teaching had gone down compared to their pre-SEI evaluations. This suggested that the students' standards were changing as a result of their exposure to research-based teaching methods. There is a fair amount of noise in this data, however, so this conclusion is not definitive.

One final caveat is that the comparison of student evaluation scores before and after SEI course transformations may be skewed by differences in class attendance. Typically, attendance is higher, often much higher, in the transformed courses than in the standard lecture courses. Hence, when student evaluations are filled out in class, the response rates are likely higher in transformed courses, which may shift the results due to a difference in selection bias of the responders. Presumably students who do attend a class that has low attendance see it more favorably than do the students who choose to not attend.

Adoption of Research-Based Instructional Practices

The cleanest and most extensive data we were able to collect were on the number of faculty who made changes in their teaching methods, and the nature of those changes. These data were provided by the regular SES reports, the annual department reports, and various faculty surveys and interviews carried out by SESs and SEI Central, and they reveal both the extent and type of changes that were implemented.

An analysis looking across the multiple sources of data shows that a substantial majority, although not all, of the changes in faculty teaching practices came about through working on SEI-supported course transformations with an SES. Almost none of the faculty adopted new teaching methods without an SES at least providing consultation or guidance in some form.

The comparison across departments as to the adoption of new teaching methods is highly informative. By combining the quantitative results given below with our extensive knowledge of the functioning of the departments and the different ways they ran their SEI efforts, we could see what factors encouraged the adoption of new methods, as well as identify a number of barriers.

CU SEI data. SEI Central at the University of Colorado conducted interviews with departmental SESs and SEI department directors in 2009, 2013, and 2014 to gather data on how courses and faculty had been impacted by the SEI. Using a structured spreadsheet, information was gathered on each course and each faculty member in the department, such as whether learning goals or clicker questions had been developed for that course, or whether a faculty member had participated in learning goal discussions or made substantial use of the SES. In the spring of 2010, a short survey was given to faculty in all the departments of CU participating in the SEI to document their level of interaction with the program and the impact they felt it had on their teaching.[2] Of the 162 faculty who were asked to participate, 114 responded. The survey responses are heavily skewed toward the faculty who were participating in SEI activities, and so we believe that few of the nonresponders had made changes in their teaching. These faculty self-report data were checked with follow-up discussions with department directors and SESs, cross-checked with annual reports from each department, then collected and coded in a massive spreadsheet showing the detailed changes that were made in all the courses and in the teaching of all of the faculty who responded. Course or faculty "impact" was defined as the total number of changes catalogued in the spreadsheet for an individual course or faculty member in terms of learning goals, assessment, and instruction, where the latter two categories are broken down into subcategories to provide a more detailed picture of the changes.

UBC SEI data. The data on course and faculty impact were somewhat easier to collect at the University of British Columbia and relied less on self-reports, as the requirement for an annual report from each department, including details on each course transformation and the faculty involved, was established from the beginning. At UBC, the changes in courses and teaching made by faculty were more likely than at CU to be part of a systematic course transformation in which faculty worked with an SES, and hence were more likely to be documented in one of these reports, which were prepared by the SESs and SEI department directors. These reports did not capture changes in teaching that faculty members might carry out on their own, for example, in response to discussions or workshops run by an SES or inspired by previous work on an SEI course transformation. However, since there were more SESs at UBC than at CU, and they were thoroughly embedded in the departments and interacted routinely with many faculty, it is unlikely that there were many such course changes that the SESs did not have some part in, although we do know of a few. We took the data from the UBC annual reports, and in some cases supplementary reports on specific course transformations, and coded them in a similar fashion as done with the CU data, analyzing them in terms of the specific changes made in the course or in the style of instruction, and using this to create a second massive spreadsheet that classified the extent of changes in course and faculty teaching across all of the SEI departments.

Quantitative Summary of Changes in Teaching by Department

In Table 5.2 we show the impact in each department according to (1) the number and fraction of faculty in a department that made major or modest changes in their teaching methods, (2) the number of courses in which teaching changed substantially, and (3) the number of student credit hours (absolute and as fraction of total provided by department) being taught in a significantly improved manner.

The most common changes in instruction were (1) adoption of learning goals that define desired outcomes in operational terms of student competencies and attitudes; (2) incorporation of various in-class active learning methods such as peer instruction with clicker questions, collaborative worksheet activities, and think-pair-share; (3) reflections on learning such as two-minute papers at the end of class; (4) new methods of assessment such as standardized pre-/post-course testing of learning each year, two-stage exams, and graded homework; and (5), pre-class reading or other activities with quizzes as preparation for upcoming class. The specific combination of practices adopted by any particular instructor varied according to individual preferences and departmental interests. To be classified as a "large change" change in instruction required the adoption of #1 and #2, and most large change cases included additional improvements. The full range of improvements is largely reflected in the list of elements on the Teaching Practices Inventory that received points for demonstrating improved learning in research studies.[3]

Number of Courses and Faculty Changed

Table 5.2 shows the total numbers of courses and faculty changed by the SEI at the University of Colorado at Boulder (CU) and the University of British Columbia (UBC).

The SEI clearly has had a substantial impact on the educational experience for the students at these two institutions. The teaching of 71 courses at CU and 164 courses at UBC has been changed as of August 2015. By the time this book is published, those numbers will be higher. In ten of the twelve SEI departments, well over half the credit hours provided by the department are now taught quite differently, and in total about 200,000 student credit hours per year (139,000 at UBC and 53,000 at CU) are now being delivered using substantially better teaching methods than before the SEI. More than 250 faculty members are teaching differently as a result of the SEI, and in seven departments this includes more than 50 percent of the regular faculty. Even in departments where only a relatively small number of faculty have made changes in their teaching and a small number of courses were changed, the teaching of a large fraction of the student credit hours being taught by the department was affected.

Reasons for Variations in Results between Departments

The best indicator of the overall impact on teaching and departmental culture is the fraction of the faculty that have made large changes in their teaching. This indicates both a willingness to consider thinking about teaching in a different way as well as learning how to actually teach differently. The fraction of department faculty that have made such changes in their teaching varies between 10 percent and 93 percent.

Table 5.2. Impacts on courses, students, and faculty teaching

CU	Impact on courses		Impact on students		Impact on faculty teaching			
	# Large change	# Significant change	# Credit hours changed (in thousands)	% Credit hours changed	# Faculty total	# Large change	# Significant change	% Change
CHEM	6	1	14.1	58	53	8	0	15
GEOL	11	7	6.3	73	32	19	5	75
MCDB	8	4	7.2	65	34	9	8	50
IPHY	7	4	9.4	72	26	12	5	65
PHYS	7	1	1.8	7	50	12	8	40*
EBIO	9	6	14.2	75	35	14	2	46*
Total	48	23	53			74	28	

UBC	# Large change	# Significant change	# Credit hours changed (in thousands)	% Credit hours changed	# Faculty total	# Changed	% Changed
EOAS	32	16	14.8	85	46[a]	43	93
PHAS	18	7	19.5	75	61	34	56
BIO	19	8	35.5	62	118	43[b]	36*
STAT	8[c]	1[c]	6.5	83	16	14[c]	88
CS	21	6	24.8	86	45	28	62
Math	5	10	20.2	38[d]	67	7	10
Other[e]	7	6	18.0			11	
Total	110	54	139.3			180	

Table 5.2. *(continued)*

Notes: "# Large change" is the number of courses in each department in which there have been major changes in the course design, instruction, and assessment, following the SEI model, while "# Significant change" means that there have been at least good learning goals adopted and there has been significant change in teaching methods, such as adding regular use of active learning in the classroom. "% credit hours changed" is the percent of the annual credit hours taught by the department that are now being taught differently. The impact on faculty section gives "# Faculty total," the total number of regular faculty in the department who regularly teach (although not necessarily regularly teach undergraduates; that breakdown is not available), and "# Changed," the number of faculty who have worked to changed their teaching in a substantial way. At UBC, in most cases this involves working with an SES to completely transform a course. At CU, there was a larger variation in the extent of changes adopted by the faculty, so these are broken out into two categories.

a. This does not include three faculty who have indicated they will retire within a few years and so were phasing out of undergraduate teaching.

b. Of the three biology departments—botany, microbiology and immunology, and zoology—there is substantial variation in the degree of involvement, with botany faculty most heavily involved and microbiology least.

c. As discussed in this chapter, a number of faculty in this department made substantial changes in their teaching with only minor consulting from SEI-supported staff, so good data on the full extent of the course transformations is not available. The actual number of courses that have been affected is higher than this. The number of transformed courses includes eight of ten courses offered by the department at the 100, 200, and 300 levels.

d. 90 percent of this impact involved adding recitation sections with guided group problem-solving and computer-graded homework to the first-year courses. Although this was a significant and beneficial change, for the courses in other departments the changes to instruction are considerably more extensive.

e. Includes four general science courses that were created in the transformed format; the remaining courses are in chemistry supported with a pilot grant.

* These departments are special cases, so this percentage does not fully represent the extent of change.

From examination of the regular reports of the SESs, discussions with SEI departmental directors and department chairs, and many conversations with faculty, the reasons for these variations can be understood, and they offer many lessons for any effort trying to bring about institutional change in undergraduate teaching.

To a large extent, these variations simply reflect the level of success at consistently implementing the three essential elements of SEI teaching transformation:

- An SES with the necessary training and disciplinary knowledge
- A faculty member willing to work collaboratively with the SES to transform a course, and in the process try new teaching methods and ideas
- A teaching assignment that has the willing faculty member (and/or collaborating faculty members) teaching the course for the necessary number of terms to successfully carry out the transformation

There have been failures with achieving each of these three, but the second was the most frequent source of problems. All the SEI departments have also found it challenging to achieve the level of planning and organization needed to have multiple SESs within a department all consistently working effectively and efficiently.

The extent to which all three elements were achieved depended on many details of how a department operates and how they ran their SEI activities. The comparisons of the different departments have been very useful for elucidating the factors that affect success. I see three natural divisions of the departments: low performing (10–15 percent faculty change), high performing (50–75 percent), and excellent (88–93 percent). In addition, there are a few special cases where these percentages do not accurately reflect their achievement.

Low Performing Departments

The low extremes, 10 percent of the faculty changed in the math department at UBC and 15 percent in the chemistry department at CU, are dramatically lower than any of the other departments. The gap is somewhat larger than it appears in the table, as the next three departments are special cases with artificially low percentages. For both UBC math and CU chemistry, the numbers are also even worse than these percentages indicate, as instructors who are not regular tenure-track faculty make up an unusually large fraction of the faculty that changed their teaching methods—four of the eight for CU chemistry, and three of the seven for UBC math.

In the case of UBC math, this failure to achieve change is clearly associated with the culture of the discipline and the department. The regular tenure-track math faculty were extremely resistant to changing their teaching methods. Most of the changes came through working with instructional faculty who were not regular tenure-track research faculty, or by adding beneficial practices that the faculty had little involvement with, such as recitation sections with active learning added to courses.

It appears that math as a discipline is highly traditional in its teaching and more resistant than other STEM fields to adopt research-based teaching methods. For example, nearly all math departments still insist on using chalk and chalkboards for all teaching; the discipline has other strongly held traditions and views about teaching and learning. The culture of the department with regard to undergraduate education is also reflected in two other observations. First, the bulk of the department teaching is in the form of large introductory service courses. The tenure-track faculty have little involvement with these courses, which are predominantly taught by graduate students and postdocs as a condition of their employment. The postdocs have little teaching experience and little incentive to teach well; and most are from foreign countries and have little familiarity with the UBC students or the educational system.

A second indicator of the unique perception of undergraduate education held by the math department comes from a survey given to all the SEI departments. In a survey of instructors asking what they believed to be the primary impediment to improved student learning, the instructors in math overwhelmingly said the main impediment was shortcomings in the students (preparation, skills, or work habits). Instructors in other departments also mentioned student shortcomings as an impediment, but far less frequently.[4] These factors suggest that there is a general view in the department (and possibly the discipline) that undergraduate education in general is not an important activity and not one where they should be investing time and effort to try to improve. It is possible that another contributing factor to the resistance to change is that math is not an empirical discipline, unlike the rest of science and engineering, and hence is less persuaded by experimental studies of teaching methods and student learning. I recognized from the beginning that it would be challenging to make progress in improving the teaching of math, but I attempted this as an experiment because the need and opportunity for improvement was so conspicuous at UBC. Also, there was a new and particularly effective chair who expressed strong commitment to the effort. Unfortunately, that person took another position not long after the SEI funding was provided to the department.

In spite of these structural elements, there have still been indications of progress. After several years of SEI support and encouragement, several regular faculty have now been making changes in their teaching methods, and a group of graduate students have become active in learning about and implementing new teaching methods.

In CU chemistry, only 15 percent of the regular instructional faculty made any changes in their teaching, and only half of those were tenure-track faculty. The failure here was again the lack of faculty willing to participate. Unlike the UBC math department, the lack of success in the CU chemistry department did not seem to be so ingrained in the discipline, but rather stemmed from problems with the general functioning and culture of this particular department. There are long-standing deep divisions within the department, and so the department struggles to come to a consensus and make a unified effort on many issues. The faculty have a host of ongoing concerns that occupy much of their time and attention, making undergraduate education a low priority. There is no position of authority within the department that is responsible for overseeing undergraduate education. Finally, the chairs turn over quickly (every two to three years), and the new chair who came in after the SEI had started was not supportive of the SEI program. He showed no hesitation in reneging on the commitments made in the department's SEI proposal that had been put forward by the previous chair.

This department was funded before we realized the need for more specific commitments in terms of which courses and faculty would be involved. Although the department had voted to support the SEI proposal, it later became clear that few faculty members were involved in the SEI planning or discussion, or were themselves willing to participate in SEI course transformation activities. While many agreed that improvement was possible and needed, few had interest in spending time on it, and the department was unwilling to provide incentives for such activities. Funding for the department was phased out, although the decision and timing were complicated by the fact the SEI was supporting major improvements underway in the large introductory courses, something we were anxious to preserve.

Although this was clearly a failure to achieve the widespread change that was the goal of the SEI, there was nevertheless a substantial benefit to students. These changes substantially improved the teaching of 58 percent of the credit hours taught by the department. That is because the department teaching load is heavily based on large introductory courses, and most of the small number of regular faculty and non-tenure-track faculty involved with those courses were enthusiastic about participating in the SEI efforts. I do worry about the long-term sustainability of the educational improvements made in that department, however, when there have been so few faculty involved.

Special Case Departments

In terms of the fraction of faculty that made improvements in their teaching, there is a large jump up to the next group of three departments, in which 36–46 percent of the faculty made substantial changes in their teaching. However, all three of these are special cases, and so a direct comparison of these percentages with the other departments underrepresents their levels of success.

The CU ecology and evolutionary biology department must be considered a success, in that it first received SEI support several years after the other departments and with a lower level of funding, but it already has 46 percent of its faculty teaching differently. Their results for both credit hours and fraction of faculty changed are impressive for such a short time, and both those numbers are continuing to increase. Looking in more detail at how the department functions and how the SEI efforts were run, this department shares most of the characteristics of the most successful SEI departments discussed below.

The CU physics department is an SEI anomaly in that, by design, the SEI effort focused on changing the teaching of only a small set of upper-division courses for majors. That is the reason it has impacted a low fraction of the credit hours relative to the other departments. The main reason for this focus is that, prior to the CU SEI, all of the department's large introductory courses had already been transformed much along the lines of the SEI model, along with the teaching of many of the faculty. Thus this particular departmental effort was intentionally quite different from other SEI efforts from the beginning. Not only was it unusual in focusing on a small set of upper-division courses, but the effort was led by a single faculty member who had substantial outside funding for physics education research connected with the effort. It is notable that an effort that targeted such a small number of upper-division courses has managed to impact as large a number of faculty members as it has. One reason for this is the unusually frequent rotation in the instructors teaching these courses compared to the frequency of rotation for upper-division courses in other departments; most upper-division science courses at CU and UBC have very little rotation and are hard to transform. Another difference in the CU physics SEI effort is that many of these faculty are using materials developed and given to them for teaching a specific course, but unlike most other SEI course and teaching changes, these faculty participated little in the design of the course transformation itself. There is evidence that this has resulted in less sustainability of the changes.[5]

The UBC biology program had 36 percent of the faculty change their teaching, but behind this number is a more complex story, largely demonstrating the importance of basic organizational structure and leadership. It was one of the first two programs funded at UBC, in part because on paper it had an established organizational structure for overseeing coordinated undergraduate education and instruction across the three biology departments. In fact, the structure existed only on paper. Instead of the three departments jointly running the program, in reality no one did. The people who taught the lower-division courses, many of them long-term sessional instructors, were left to do as they pleased, and no one felt able to exert any authority over them, particularly as there was such a long-established precedent for not exercising any supervision or authority. This was particularly problematic because most of these courses were multiple-section, multiple-instructor courses, with each instructor acting independently, even to the extent of covering their own chosen set of topics and giving their own exams. These structural problems resulted in SESs struggling to work with these instructors but making very little progress, and ultimately quitting to take other jobs. While the department chairs expressed concern and a desire to change things, they felt unable to do so within the existing structure. In response to these problems, we greatly reduced SEI funding to the biology program.

Over the subsequent few years, with considerable input from the dean, the organizational structure of the biology undergraduate program was changed and good people were put in positions that now had clear responsibility and authority. They developed a clear plan for the development/transformation of a set of courses that would reshape the biology curriculum, including identifying the faculty members who would have responsibility for laying out what would be taught in those courses and the pedagogy used, aligned with the SEI goals. In response, we reinstated their funding, and since that time the progress has been good. They have systematically changed the curriculum and teaching methods of many large courses, which now provide more than 35,000 credit hours per year and involve forty-three faculty members. The process is coupled to a curriculum reform, which involved a shift in responsibilities of the three participating biology departments. While this reform complicates the SEI work in some respects, it also has benefits in making it part of a larger effort. Although at 36 percent the fraction of the faculty that have changed their teaching appears low, this is slightly misleading, as that fraction is the total across all three biology departments, but the botany department is now taking a larger responsibility for undergraduate education within the new alignment and has a large fraction of faculty involved, while the microbiology department has relatively little responsibility and few faculty involved.

The biology program has provided a dramatic example of how the organizational structure by which an undergraduate program is run can have a large impact on the quality of the program and how it can (or cannot) innovate and thrive.

High Performing Departments

These departments have had 50 percent or more of their faculty change their teaching, and two-thirds or more of the credit hours provided by the department are being taught using research-based methods following the SEI model. Within each of these departments are a variety of different situations that affected the degree of success and explain why they are not at the 90 percent level of the most successful SEI department. All have had some difficulties with departmental planning and management of the SEI efforts, and all have little rotation of teaching assignments among upper-division courses. This lack of rotation leaves some faculty with teaching loads dominated by the teaching of one or two upper-division courses, which they are seen as "owning." As discussed below, this pattern of teaching assignments can leave such faculty quite isolated from broader considerations and interest in undergraduate education, which limits the fraction of faculty impacted by the SEI. In addition to these common features, each department has some special challenges of its own.

UBC physics and astronomy (PHAS) had a particularly problematic organizational structure in which historically essentially everything was done by the chair with little delegation, including the running of the department's SEI effort. For such a large and varied department, even without the SEI this is an impossibly difficult job. It meant the attention devoted to the planning and structure of SEI activities and to oversight and guidance of the SESs was inadequate. This resulted in a substantial amount of SES time (and hence SEI funding) being used ineffectively. A special challenge for this department was that it focused much of its effort on changing the large introductory courses, which, like in the biology program, are multiple-section, multiple-instructor courses where the instructors have become accustomed to acting independently. Although it involved substantial work and several false starts due to insufficient planning and oversight (not ensuring that all the involved instructors were either committed to the effort or replaced), eventually PHAS was successful in transforming these courses. They have now established a common mode of quality instruction and content, which new faculty now rotate into and adopt. This is a major accomplishment.

CU molecular, cellular, and developmental biology (MCDB) has achieved changes by 50 percent of its faculty, impacting 65 percent of the credit hours. The main limitation on achieving wider impact within this department is that teaching loads are relatively light, and many of the tenure-track research faculty primarily teach only their particular upper-division specialty courses. As noted, such courses and faculty are particularly difficult to change. This department's SEI efforts have also been limited by a uniquely difficult personnel conflict, which tends to disrupt any attempt to arrive at departmental consensus and actions, particularly when teaching is involved. This personnel conflict is also an ongoing distraction to the chair, dominating the time and attention that the chair can put into the SEI and taking away from more constructive activities.

CU integrative physiology (IPHY) has been quite successful, at 65 percent of the faculty changed and 72 percent of the credit hours. The department had a supportive chair and receptive faculty. It likely could have achieved a larger impact among the faculty if the SEI departmental directors had been more aggressive about planning course transformations and recruiting and incentivizing faculty to participate, particularly those who primarily teach upper-division courses.

UBC computer science (CS), with 62 percent of the faculty and 86 percent of the credit hours changed, has been successful, but CS has followed an approach rather different from the other departments. They have had much more difficulty hiring SESs than other departments, because of the strong industry competition for people with CS skills. They also struggled for several years with departmental leadership. The chair had difficulty getting the faculty to work together, with the desires of individual faculty members taking precedence over departmental plans and commitments on various educational changes and planned SEI activities. Although there was some initial progress, it was relatively slow. A significant early accomplishment was the creation of learning goals for their first- and second-year core courses. After some changes in leadership, they ended up with an effective and committed chair who worked productively with the SEI department director, and they solved the SES problem.

The solution was to use a different model, with much of the SES activities done by tenure-track teaching faculty who historically have played a large role in the department and are well respected. Using SEI funds, the department bought out some of the teaching time of these instructors so that they had time for more training about science education research and for serving as SES consultants to the rest of the department. These SES instructors also took the lead in establishing learning goals for the main academic streams of CS majors and mapping these goals onto the courses in that stream. This generated discussion with many faculty as to the educational goals of these tracks and the courses involved in them, and how well these goals were being met. CS was also different from other SEI departments in that it used a larger fraction of its SEI money to support many smaller teaching projects that individual faculty would propose and carry out with SES consultation and advice, rather than supporting full scale course transformations. Through this process many faculty have adopted new teaching methods and many SEI course elements, without extensive course transformations supported by an SES.

In *CU geology,* 75 percent of the faculty have changed their teaching, and 73 percent of the credit hours have been impacted. This department had a supportive chair and a receptive faculty, with some strong faculty proponents. A factor that likely helped was that nearly all of the faculty in the department cycle through teaching the two large introductory courses, which provide a large fraction of the department's credit hours; SESs could reach most of the faculty through these courses. One area for possible improvement would have been better training of the SESs. In the most successful departments, we saw that the SESs would find particular opportunities for instructional change that would result in immediate and obvious improvements in areas of concern to the instructors. This would convince the instructors of the value of these research-based teaching methods, and they would talk about them to their colleagues. In CU geology, however, the early changes resulted in less obvious improvements and had little emotional impact on instructors. I worry that this may have affected the willingness of faculty to sustain and build on instructional changes in the future. In later SES training, we added an emphasis on the need to learn the instructors' concerns and find interventions that would directly target them, but CU geology was the first department to be funded and to hire SESs, and at that time the SES training program was immature.

Excellent Performers

UBC statistics is a very small department with a correspondingly low level of SEI support, but which they have used to good effect. They have a lot of rotation in teaching assignments, with nearly all faculty teaching the courses that provide most of their credit hours. Thus nearly 90 percent of their faculty have changed their teaching. Also, in addition to SES-driven efforts to make changes in courses, there have been a few senior faculty who made major changes in their teaching, after discussion with other faculty and SESs, but with little direct SES support. The result has been a general overall change in how the faculty in the department teach and talk about teaching. Few upper-division courses have been changed as an SEI effort, but there may be faculty who have changed how they teach those and we are not aware of it because of the limited SES presence and reporting in the department.

The UBC Department of Earth, Ocean, and Atmospheric Sciences (EOAS) was clearly the most successful large department at achieving widespread improvement in their teaching. Nearly all of its courses have been transformed and nearly all of its faculty have adopted new teaching methods. The level of success enjoyed by this department deserves a closer investigation, which I take up in the following section.

What Made the UBC EOAS Department So Successful

The following is a description of the set of elements that were put in place as part of the EOAS SEI effort. I see these elements as providing a model for success in any program that has goals similar to the SEI's for improving teaching.

Teaching Initiatives Committee. A new, permanent Teaching Initiatives Committee (TIC) was established to plan, coordinate, and guide the SEI program. The chair of this committee became the department's SEI director. (Note that this is not the curriculum committee; the TIC is tasked with overseeing teaching initiatives within the department and does not have the role of approving curriculum changes.) In addition to the SESs and the SEI department director, the committee includes two or three other long-term faculty members, and usually graduate student/TA and undergraduate representatives. The TIC provided a good guidance structure and the SEI department director led the program in a very competent, organized, and thoughtful way. Later in the program the department director took a one-year sabbatical, but the program was well established by that time and was capably managed by long-term SESs during the director's absence.

Consistent strong leadership and support. Although the department chair changed twice during the SEI program, all of the chairs were very supportive of the SEI efforts. Among many other supportive actions, the chair made it explicit that if faculty student evaluations go down during the course transformation process, the chair would take responsibility for contextualizing those evaluations in promotion and tenure processes. (In reality, evaluations usually stayed about the same.) The chair also often put items relating to the EOAS SEI on the agenda of faculty meetings and retreats.

Detailed planning. A detailed plan was developed by the TIC, identifying which courses and faculty would be involved in transformations and what the timeline would be. This ambitious plan was designed to involve as many faculty as possible working with SESs in an efficient manner. It served as an overall guide for the program, listing which courses would be transformed, and breaking the transformations down into planning, teaching, and second teaching terms. The plan evolved during the program but kept its original scope and intent. The plan as of January 2009 is shown in Figure 5.1.[6] One factor that was different in EOAS compared to the other science and math departments at UBC and may have simplified its planning process is that it does not have any large required service courses (i.e., courses that are required for students majoring in other disciplines). EOAS does have a large elective course (EOSC 114) that was successfully transformed early in the program. It has an enrollment of about 650 per term and has multiple sections and sequential instructors, with one instructor clearly in charge, which helped make the process go smoothly.

Science education specialists. All of the initial SES hires were internal people (former students, postdocs, lecturers). This was not the explicit plan, but it turned out that the best candidates were internal. This meant that the SESs were familiar with the department, and a subset of the faculty was familiar with each of them. Thus it was relatively easy to begin comfortable conversations about teaching, and get the course transformations started relatively quickly.

The EOAS SESs were all hired into temporary faculty positions. They attended faculty meetings and frequently participated in discussions about teaching and learning at those meetings. Two of them stayed on as SESs for seven years, each working with more than fifteen faculty members. At the end of SEI funding, these SESs continue to be employed in EOAS, and a number of EOAS faculty members continue to consult them on teaching issues. There were also a number of other SESs hired; during the middle four years of the program, there were typically four SESs working in the department, each working with two to four faculty members.

While this department has been quite successful with internal people becoming SESs, we do not feel this is necessary for a successful SEI department. In looking across the SEI departments as a whole, we have found that external SESs can also be very successful, but it usually takes longer for them to become familiar with a department and develop good working relationships. On the other hand, external people can bring experience and perspectives that might be lacking internally.

Direct incentives to faculty members. At EOAS, the SEI had a menu of possible incentives to faculty in order to get them to put in the work necessary to transform a course. This included, for each of three terms (one planning term plus two teaching terms), either (1) a 0.5-course release, (2) a six-hour-per-week extra TA, or (3) something equivalent that would take work off the faculty member's plate, such as partial support for a research assistant.

As of 26 January 2009

EOS-SEI LONG-TERM PLAN

UPDATED DRAFT, STI LL FLEXIBLE

P1 = first planning term; P2 = second planning term; T1 = first teaching term. etc.

TARGETED COURSES	2007	2008			2009			2010			2011		
	Fall07	Spr08	Sum08	Fall08	Spr09	Sum09	Fall09	Spr10	Sum10	Fall10	Spr11	Sum11	Fall11
EOSC 114	P2&T1	P3&T2	P3	T3	T4								
EOSC 111	P2&T1	P3&T2	P3	T3	T4								
EOSC 221	P1	T1	P2	P2	T2	P3	P3	T3					
EOSC 324	*MLB*												
ENVR 200	*DS&SH*												
EOSC 112		P1	P1	P2&T1	P3&T2	P3	T3	T4					
EOSC 220		P1	P1	T1	P2	P2	T2	P3	P3	T3			
EOSC 212		P1	P1	T1	P2	P2	T2	P3	P3	T3			
EOSC 210		P1	P1	T1	P2	P2	T2	P3	P3	T3			
EOSC 116		*SS*											
ENVR 300		*DS&KC*											
332 (JM)				P1	T1	P2	P2	T2	P3	P3	T3		
322 (GO)				P1	T1	P2	P2	T2	P3	T3			
355 (CJ)				P1	T1	P2	P2	T2	P3	T3			
EOSC 449				*MLB*									
ENVR 449				*KO*									
ATSC 201				*RS*									
EOSC 211 (RP)					P1	P1	T1	P2	P2	T2	P3	P3	T3
EOSC 372 (SA)					P1	P1	T1	P2	P2	T2	P3	P3	T3
EOSC 373 (MM/others)							P1	T1	P2	P2	T2	P3	P3
EOSC 252 (FH)							P1	T1	P2	P2	T2	P3	P3
EOSC 472 (KO)							P1	T1	P2	P2	T2	P3	P3
EOSC 321 (MK)								P1	P1	T1	P2	P2	T2
EOSC 331 (KH)								P1	P1	T1	P2	P2	T2
EOSC 326 (SS)								P1	P1	T1	P2	P2	T2
EOSC 329 (RB)								P1	P1	T1	P2	P2	T2
EOSC 222 (PS)									P1	P1	T1	P2	P2

Courses undergoing transformation w/o specific STLF help

Course sequence considers: logical progressions, breadth in EOS, faculty keenness

" *aims for: maximum departmental involvement*

" *interfaces with: teaching assignments, scheduling, & sabbaticals*

FIGURE 5.1. The EOAS-SEI long-term plan

Note: Name was changed from EOS to EOAS partway through the SEI.

Table 5.3. EOAS course transformation expectations agreement

	By end of planning term	By end of first teaching term	By end of second teaching term
Project scope	Outlined	Revised	In final documentation
Course-level learning goals	Draft: involve stakeholders	Revised	Broadly accepted
Module- or lecture-level learning goals	Draft	Revised	Mapped to course learning goals
Assessment	Draft plan	Revised plan and materials	Optimized plan and materials
Teaching methods (pedagogy)	Draft plan	Revised plan and materials	Optimized
Short summary of structure and rationale		Draft	In final documentation
Materials archived			Completed
Plan for sustainability			Completed
Share progress/problems	Annual or semiannual mini-retreat		

Agreement on responsibilities. An agreement was also developed to give faculty a detailed understanding of what would be expected when they participated in a course transformation. This agreement was signed by the faculty member and the department chair. Table 5.3 shows the expected elements for each of the terms of a course transformation as laid out in the agreement.[7]

Progress reports and regular meetings to discuss progress and strategies. The EOAS SEI department director would meet with the SESs weekly, regularly providing suggestions, offering guidance, and monitoring progress. Regular meetings between the department director and SESs happened in other UBC SEI departments as well, but the EOAS meetings were particularly focused and well aligned with advancing the departmental

SEI program goals. In addition, short written progress reports (initially twice per month and later monthly) were also required of all the SESs. These were typically discussed at a monthly meeting that included the SESs, the SEI department director, and the UBC SEI director.

Teaching assistants' development. The EOAS SEI developed a course for graduate TAs: EOSC 516, Teaching and Learning in Earth and Ocean Sciences. The course was designed to improve the teaching skills and knowledge of effective pedagogy of the TAs. The course is now run primarily by graduate students who have facilitator training, and has an enrollment of about fifteen students per year.[8]

Involving undergraduate and graduate students. In addition to the involvement of graduate students in the running of the TA training course described above, twenty graduate students were supported by SEI funding to be involved in improving courses under the SEI, ranging from redesigning laboratory courses to developing learning goals and in-class activities. Ten undergraduate students were involved in various aspects of the EOAS-SEI, and three of them completed geology honors theses based on their SEI-related educational research.

Communication within the department. The department as a whole discussed and approved each of the key planning and implementation documents used in the project—for example, the long-term plan, the incentive agreement, and the course transformation expectations agreement. Thus these key components weren't implemented without an opportunity for everyone to have input. Broader departmental involvement was fostered via brown-bag seminars on education (which had only modest attendance), invited speakers in geoscience education research as part of the normal departmental seminar series, regular SEI postings on the departmental bulletin board, and the *EOAS-SEI Times* newsletter. The monthly newsletter reported on SEI accomplishments and progress and was put into faculty mailboxes and posted on the departmental website.[9] Seminar topics included discussion of effective clicker use, attitudinal survey results, midterm and end-of-term surveys, improving exam questions, online discussions as a learning tool, critical incident questionnaires, assessing geoscience programs, and just-in-time teaching. The EOAS SEI program also maintained a website that gave details of the projects being done under the SEI as well as resources for faculty members.[10] Finally, some respected senior faculty became conspicuously involved with the SEI early on.

For several years the department had an annual SEI mini-retreat in April. During that half-day event, all the EOAS instructors currently involved in SEI would meet to share and discuss what they had been doing in their courses. This allowed a space for busy faculty members to talk to and learn from one another about teaching, which seldom happened spontaneously during the academic terms and across the many subdisciplines in EOAS. However, by the end of the SEI such spontaneous discussions had become far more common than they were at the start. The SEI work was also an explicit topic of discussion at the annual departmental one-day retreat for the first four years of the project, and again during the year of transition to the post-SEI era.

In addition to other resources provided to the faculty, SES Francis Jones spearheaded the creation of the Evidence-Based Science Education in Action video series.[11] These professional videos show a variety of innovative teaching methods in use in real EOAS, math, and physics classes. The videos came with supporting materials to provide context, instructor's tips, and pertinent references.

Each of these ten components listed contributed to the success of the SEI effort in EOAS Rather than any single critical element, I believe it was the thoughtful combination of all of these elements that is unique to this department that made this department's SEI efforts so successful. The success was as a result of having committed people in positions of authority who understood how to manage organizations and the people involved.

Teaching Practices Inventory Data for Departments

Another source of data on the teaching changes accomplished at UBC by the SEI is provided by the Teaching Practices Inventory. TPI data exists from the UBC SEI departments for the 2012–2013 year.[12] In keeping with the challenges discussed in Chapter 3 about getting departments to collect data, it was difficult to convince departments to require faculty to fill out the TPI survey. For UBC, only EOAS and CS set the expectation that all faculty should do this, and hence obtained a sufficiently high compliance rate (about 90 percent). For CU, we received only a handful of responses. Only for UBC EOAS do we have adequate data for both 2006–2007, when

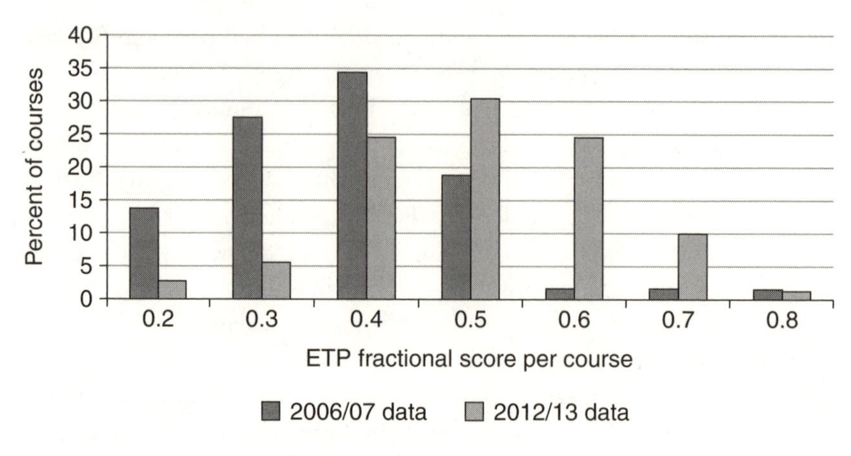

FIGURE 5.2. ETP scores for courses in EOAS

This histogram shows the fractional ETP scores for the courses in the UBC EOAS department in the 2006–2007 and 2012–2013 academic years. The survey was slightly different for the two dates, so the scoring is the fraction of the maximum possible score based on the subset of forty scored questions common to both versions of the inventory. (See note 3 in this chapter.)

the SEI was just beginning, and 2012–2013, so we can examine the change. As shown in Figure 5.2, there is a substantial increase in the TPI scores, representing a substantial increase in the extent of use of research-based teaching practices (ETP). The comparison between CS and EOAS 2012–2013 TPI scores shows that they are similar overall, although a more detailed analysis of the different categories shows more variations. The overall similarity is consistent with Table 5.2 showing that there have been changes in teaching in both departments for a large fraction of their courses and credit hours.

Sustainability

While it's unclear whether the transformations carried out in the courses and the changes in teaching methods of individual faculty members will be sustained, there is short-term data on this from the departmental annual reports and some surveying of the faculty. These indicate a high level of sustainability at UBC. A 2013 survey of the seventy faculty members who had adopted substantial changes in how they teach as part of the UBC SEI program and then had at least one subsequent year teaching without any SEI support showed that all but one of the seventy had continued to use the new methods they had adopted.[13]

Furthermore, in that same survey, 90 percent of the faculty who subsequently taught a different course without SEI support reported that they had adopted some or all of these novel teaching methods in that subsequent course.

There is some indication from the CU physics department that faculty who adopt methods and materials to teach a transformed course but never actively participate in the design process for transforming a course are less likely to sustain the use of new teaching practices.[14] However, there are also a few examples of faculty getting a brief and relatively superficial exposure to new teaching methods, but then over the course of months or years embracing them more and taking time to understand and use them effectively. To truly know the extent of sustainability of the SEI impact on teaching, it will be necessary to wait and watch.

I suspect that the fraction of the faculty—particularly the regular tenure-track faculty—that have adopted research-based teaching methods in a department will likely be a good predictor of departmental sustainability. I am concerned about sustainability in those departments where fewer than 50 percent of the faculty members have adopted new teaching methods, even if the teaching of most of the credit hours has changed. As noted, there are several departments where a significant fraction of the faculty teach only specialized upper-division courses and have not made any changes. Although they may teach relatively few credit hours and in limited contexts, those faculty members still speak with an equal voice in hiring and promotion decisions and discussions about how teaching should be evaluated and rewarded. Departmental decisions on such issues will have long-term impacts on the methods and quality of teaching in a department. The smaller SEI program at CU has generally resulted in smaller changes in the departmental practices at CU than at UBC. I believe that this difference is likely to be reflected in differences in the sustainability of the improved teaching methods.

I know of one department (not surprisingly, UBC EOAS) that is making changes in the department's methods of evaluating teaching for merit, promotion, and tenure that are based on the department's SEI experiences. That suggests that sustainability of the use of new teaching practices in this department is very likely.

Faculty Attitudes about Teaching

Over the course of the SEI I learned a great deal about faculty and department attitudes about teaching and learning and saw many of these shift over time. Here I list the most notable observations. These are my personal impressions, but they are shaped by hundreds of conversations with faculty members, department chairs, SESs, as well as reviewing of large numbers of SES and departmental reports. I have become convinced that virtually all faculty want to teach well. I found that most faculty who use methods that are less than optimal may care as much about teaching as others do, but they are unconvinced of the value of changing. When they pursued actions counter to what we desired, there was no maliciousness in their actions; rather, it was the result of having different values and priorities, largely as a result of the incentive system and the culture in which they were working.

I also saw that nearly all faculty members can learn to use new teaching methods effectively, but there is a significant initial learning curve during which the faculty are learning what this form of teaching looks and feels like in their own class, as well as developing an understanding of the theory of learning on which it is based. While there were obvious variations in the speed and effectiveness with which faculty adopted the new teaching methods, the great majority became reasonably effective after working with a well-trained SES. In this regard, the SESs acted as coaches, sitting in on classes and regularly offering specific feedback and advice. This assistance was the most critical in helping faculty tackle the initial part of the learning curve. After that most faculty could function well and continue to improve on their own. The variations in the steepness of the learning curve among different faculty members could be largely explained by how knowledgeable they were as to the thinking of the students in their courses. The range of faculty attitudes about the adoption of teaching innovations that I observed has some agreement and some disagreement with prevailing wisdom. While faculty can be fairly well described by the general categories put forth by Rogers for adoption of innovations—early adopters, the thoughtful majority, and laggards—I found the distinctions to be rather fluid and time dependent.

Early adopters. These are the faculty members who were knowledgeable about discipline-based education research (DBER) and already implemented many of the ideas, or who had been reflecting deeply about teaching and learning and were increasingly dissatisfied with the results of their traditional methods of instruction. They saw the teaching methods espoused by the SEI as the solution they had been looking for. These faculty members often immediately and effectively put research-based teaching methods into practice. They valued the prospect of having an SES to partner with in this work. With such faculty members, SES work could focus on supporting the instructor as they incorporated the new teaching practices, and provide feedback on materials and implementation to allow for iterative improvement. These faculty members could also be cultivated as educational leaders within the department, speaking about their experiences and satisfaction.

One caveat is that we have found that a significant fraction of this cohort also like to emphasize the enormous (and usually unnecessary) amount of time they spent on their teaching. This is presumably an attempt to get greater credit and respect for their teaching efforts, but it can serve to discourage others from adopting new teaching methods. A second caveat is that it was not unusual for faculty members' stated beliefs to be inconsistent with their subsequent actions. For example, some instructors who portrayed themselves as early adopters were limited in what changes they would consider, and some self-anointed traditionalists turned out to be rather flexible and adopted new methods, sometimes even while continuing to label themselves as traditionalists.

Thoughtful majority. This group comprises the largest number of faculty members. The members of this group were not immediately convinced they should change their practices, or more specifically, that they should put in the time required to change. Usually they were concerned about the impact this would have on their research and were not sure the benefits were sufficient to offset that cost, but they were open to arguments to the contrary. They simply display the healthy skepticism with which scientists would be expected to treat any new claim.

Over a period of time the views of many of these faculty evolved as they were exposed to new ideas about teaching and learning and to data on results, whether through discussions in faculty meetings, hallway conversations with early adopters and other participating faculty and SESs, seminars, or articles and newsletters distributed within the department.

After such exposure, the typical next step was these faculty members would talk with an SES about possible changes to their teaching and/or observe a course being taught using novel methods. This was often the strongest argument for convincing faculty to change their teaching methods—seeing students far more engaged and interested in the material and asking many more, and deeper, questions than in their regular lecture classes. Observations of a class also helped reduce the common fear that allowing students to talk together to solve problems would lead to a loss of control of the class. Faculty were also able to see that substantial material could still be covered in a course taught using active methods, addressing a second common concern.

It was typical for many of these thoughtful skeptics to take a year or even two after a department had launched a full-scale SEI effort before they came forward saying that they would like to work on transforming their teaching. We learned not to prejudge which faculty members would be the most likely to support and participate in the SEI efforts, as many individuals defied initial expectations. After being exposed to SEI methods during this one- to two-year incubation period many decided to change, including some who had been thought by their department to be hopelessly traditional. The relationship of seniority to attitude change was also more complex than is often assumed. Although younger faculty may have been slightly more likely to buy into new teaching methods, there were some young faculty members who were strongly opposed to the basic concept, and many others felt this was something they should avoid until after they had tenure. On the other hand, numerous senior faculty members became very enthusiastic about the SEI methods.

Laggards. There are many references to this type of faculty member in the educational change literature, usually with comments to the effect that death or retirement is the only way to deal with them. I believe that such sentiments are too pessimistic, and that it is more appropriate to think of most of these faculty as simply the tail of the distribution of the thoughtful majority. In the most successful SEI departments, a number of faculty who had previously been seen as completely resistant to change eventually sought out SESs to help them with transformation of their teaching. In a few cases, there have been suggestions that student complaints about how much less they were learning in traditionally taught courses, compared to the transformed courses, might be a contributing factor. Educational innovations across the department can lead to such complaints, as students become accustomed to more effective teaching methods.

I do not believe that it is realistic to expect all of the laggards to change their teaching in response to an SEI type program, but I do think it is dangerous to prejudge how faculty members will respond based on their initial reactions and behavior. The reasons this group of resistant faculty act as they do are quite varied. Some have been recognized as good teachers on the basis of teaching awards given for their lecture performances and see lecturing as core to their identity as a faculty member; some feel they could never excite students in the subject and be effective teachers; and others see their real job as doing research, with teaching merely a minor side annoyance. Over the course of a few years, we have seen large changes in all of those attitudes, but not in every case.

So, while it can be useful to recognize the values and perspectives that are reflected in these conventional categories of adopters of innovation, care must be taken not to jump to conclusions about what faculty members will and will not do and why, based on how one has classified them. The beliefs and behaviors of individuals are affected by various formal and informal incentives and experiences, and these beliefs change over time scales of a year.

Institutional Contrasts in Attitudes about Long-Term SEI Impact

CU attitudes. As part of a small NSF funded external evaluation of the SEI, a CU researcher not involved in the SEI conducted in-depth, semi-structured interviews with samples of individuals involved in the change initiative: SEI leadership (including institutional administrators), project leaders, department directors, SESs, SEI-engaged faculty, and the chairs of SEI-participating departments. Interview protocols explored individuals' knowledge of the change effort, their role within it, their experiences in SEI, their attitudes and beliefs about teaching and learning, self-reported changes in these as a result of involvement in the change initiative, issues of autonomy, motivation, and resistance to SEI, and whether the initiative was seen as successful, and why or why not.

Individuals were solicited to participate in an interview, and those who agreed provided a letter of consent for the study via university email. Out of sixty-five individuals invited, fifty-four were interviewed. Interviews were conducted individually, lasted one to two hours, and were digitally recorded and transcribed. These data were released to two members of the SEI team, including myself, under a separate IRB protocol, dependent upon individual consent, with the provision that (1) the individuals would remain unidentified,

(2) administrators and former or current SESs employed at CU were not included, and (3) individuals were allowed to redact their statements. With these restrictions, a total of twenty-four agreed to release their transcripts to me.

The views reflected in these interviews were very consistent with my impressions about general CU departmental attitudes formed from previous discussions with CU SESs, faculty, and department directors. While most of the individuals interviewed were enthusiastic about the changes that had taken place in their departments, the great majority expressed the opinion that these new approaches to teaching and their benefits were not embraced by their department as a whole. Most communicated concerns about sustaining and building on these changes after the end of the SEI funding. There were also many comments expressing the general belief that the only thing that mattered in the department was research productivity, and after the end of SEI funding this emphasis would overwhelm any attention to teaching and education. It should be noted that these interviews and other sources of input did not involve the CU ecology and evolutionary biology department, as its SEI activities began rather late and appeared to be taking a somewhat different path from the other CU departments, with potentially more positive attitudes.

While there were many negative attitudes expressed about the sustainability of the SEI impact in CU departments, they may not entirely reflect the reality. In discussions with me and other members of CU SEI Central, some faculty members have expressed a lack of enthusiasm for the SEI concept, but then mention that they have adopted and planned to continue using a number of the teaching methods advocated by the SEI. Also, as SEI funding approached its end, all of the SEI departments expressed the desire to find some way to preserve an SES in the department, as they were seen to be of great value. One of the institutional differences was that this idea of preserving some SESs was not supported by the CU dean, whereas it was supported and ultimately funded by the UBC dean.

UBC attitudes. At UBC, our indications of departmental attitudes come from interviews with department chairs, SEI department directors, faculty, and, most of all, regular feedback from the SESs. While there is a spectrum of opinions, overall the views of the SEI and its long-term impact on departments are considerably more positive than at CU. (The comments below apply only to the non-math departments.) In the early years, the attitudes were similar to those expressed in the interviews at CU, but that changed over time. While there is now the general opinion that there are some faculty members who will probably never change their teaching or beliefs, the general attitude in the non-math UBC departments is that such faculty are now the exception rather than the norm. In the SEI departments there are now many faculty, including some highly respected ones, who regularly discuss the benefits and pleasures of teaching in these new ways, and a steadily increasing number of faculty who are embracing new teaching methods and seeking out help with their use. In large part because so many faculty members are so enthusiastic about these teaching methods and have colleagues around them who feel the same way, many in the departments are quite optimistic that these changes have become the norm within the department, even if they will not be used by everyone.

There are also other signs that this transformed teaching is becoming entrenched within the culture of the UBC departments. In most departments there are now ongoing discussions or established plans as to how new faculty coming into the department will be trained in the use of these teaching methods. A co-teaching program has been established in two large departments (EOAS, PHAS) in which funds are provided to support a faculty member highly experienced in these teaching methods to co-teach with a new faculty member (or in some cases senior faculty members) in order to develop their teaching expertise. The dean has recently established a program to fund a permanent SES-type position in each department, with the intention that these individuals will serve as expert consultants to faculty. This suggests a rather fundamental change in thinking about teaching, namely that it is an activity that involves true expertise that comes from knowledge and careful practice, rather than merely a matter of individual opinion and expression.

I believe that the reasons for these differences in attitudes at the two institutions are likely some combination of three factors. The first is the strong and conspicuous support by the UBC dean of sciences. Both within and outside the institution, the dean regularly discussed the SEI and what it was accomplishing, characterizing it as something to be proud of. He brought it up in his regular meetings with the department chairs, and when there were problems with a department's SEI work, he would discuss it with the chair. In his selection of new chairs, the candidate's support for the SEI activities was a significant factor, and so each new chair who came in was usually highly supportive and an effective leader. At the annual SEI mini-conference, the dean and many of his associate deans were always conspicuously seated in the middle front of the auditorium, and he was very engaged, frequently asking questions of the speakers. It was also a fund-raising priority for the dean. At CU, none of these things happened, as the dean largely ignored the SEI, and most new department chairs at CU that were appointed during the SEI were neutral or opposed to its activities.

The second reason is money. The UBC SEI had about twice as much money as at CU. This meant that UBC departments had more money to use for hiring SESs and providing incentives to faculty to participate in SEI course transformations.

The third factor is better training of SESs and better management of the SEI in general. As noted above, the UBC SESs had more formalized and more extensive initial and ongoing training. There were also more of them and they had a much stronger sense of community and used this to enhance their knowledge and skills. Also, as I will discuss at length in Chapter 6, I learned a great deal about the changes needed in how the SEI functioned, and there was more opportunity to implement these changes at UBC than at CU.

Attitudes about Learning Goals: Contrasts between Institutions

The attitudes about learning goals offer a notable contrast between the two institutions. The extent of acceptance of learning goals for structuring and guiding courses and teaching varied considerably across departments at both institutions, but over time there has emerged an overall institutional difference. Learning goals are now widely accepted as the norm within most departments of the UBC Faculty of Science, but much less so in the science departments at CU. It has become routine for UBC faculty members to discuss courses in terms of the learning goals they desire to achieve, and to start the design of new courses with identification of learning goals, even when those efforts are not connected with the SEI. It is considerably less likely at CU for there to be good learning goals that instructors embrace and use. (The CU ecology and evolutionary biology department is a notable exception to this and in several other regards.)

This difference in attitudes is a large change from the start of the initiative, when there was considerable fear and discomfort expressed about the idea of having learning goals; discomforts that were nearly identical at both CU and UBC. The faculty had difficulty in articulating good learning goals, they felt that they would be too constrained by having learning goals, that showing goals to the students would result in complaints about the teaching and the exams, or that having such goals would necessarily make the courses more superficial. It should be pointed out that these common fears about learning goals are entirely in the abstract. I am not aware of any of them ever actually occurring in the hundreds of SEI courses that have learning goals that are available to the students.

I am not sure why the attitudes evolved differently at the two institutions, but I speculate that there are two main reasons. The first is that the dean at UBC mandated that there had to be learning goals for all the introductory courses that satisfy a college or university requirement (which includes courses in math, physics, biology, and chemistry). Second, there were more SESs in the UBC departments to assist with the formulation of good learning goals, and at UBC they were better trained on this than the SESs at CU. These differences resulted in more faculty actually going through the process of creating goals and using them in their courses. This resulted in their seeing the learning goals as more familiar and less threatening, and ultimately as beneficial.

It is interesting to note that, unlike UBC, there is an accreditation process at CU that requires such learning goals for all courses. However, for accreditation purposes usually one individual in the department creates and turns in to the appropriate office the goals for the courses, working largely in isolation. It is unclear whether this process causes the faculty to be more cynical and suspicious about learning goals, or whether most are simply oblivious to this part of the accreditation process.

While the original SEI vision was to create learning goals that reflected a departmental process and consensus, this almost never happened. I think this was because it was simply too difficult and unfamiliar a task and that it involved too much collective effort to be worth the perceived benefit. However, it is likely that many sets of learning goals that were created by an individual faculty member for their course will end up accepted as the de facto departmental goals. Whether future instructors of the course will use those goals to guide how they teach and assess students, and whether departments will monitor if that is happening, is unclear.

Economics of the SEI: Ongoing Costs, One-Time Costs, and Private Fund-Raising

In addition to tracking the impact of the SEI on teaching and departmental attitudes, we also collected data on instructional costs after the completion or near-completion of the SEI transformation efforts, and compared those to the pre-SEI costs. These data confirm the assumption of the SEI model, which is that providing more effective undergraduate instruction costs no more than traditional lecture instruction.

We also provide a brief analysis of the return on the SEI investment per instructional credit hour transformed, although that return was not a goal of the program. Finally, we include a short note here on our experiences with regard to fund-raising for an SEI-like enterprise.

Ongoing Instructional Costs after the SEI

Costs at UBC. About 180 faculty members significantly changed their teaching practices, changing the teaching of 140,000 credit hours per year. The original design of the SEI was that there would be substantial one-time transition costs, but that at the completion of that transition, the instructional costs would remain the same. These costs include the faculty and administrative salaries associated with the undergraduate courses and the cost of TAs. We have examined the changes in those costs as a result of the SEI activities. We have excluded the normal enrollment-driven adjustments from this analysis, as those are disconnected from the SEI activities. The changes in costs for UBC are listed in Table 5.4.

There was no change in the number and cost of instructional faculty or administrative support across any of these departments. There was no change in the number of TAs in statistics, math, computer science, or

Table 5.4. Changes in instructional costs at UBC

Department	Change in faculty and administrative costs	Change in TA cost
EOAS	None	Added training; increased cost about 2 TAs=$12,500 per yr
PHAS	None	Added training and numbers; increased cost about 4 TAs=$25,000/yr
Statistics, math, computer science	None	None
Biology	None	None (although reallocation)

biology, and hence no change in ongoing costs in those three departments. There was some small reallocation of TAs, largely to be somewhat more rational. For example, in biology it was realized that the ratio of TAs per credit hour was roughly twenty-five times higher for upper-division courses than for lower division courses, and so there was a small reallocation of TAs to lower division courses as SEI transformations were carried out.

EOAS and PHAS both introduced TA training programs. Led by mentor TAs, these programs are run in the week before the term starts and have some follow-up during the term. This training costs the equivalent of two TAs per year. PHAS also increased the number of TAs assigned to introductory courses by two to help manage the logistics. So, the net increase in instructional costs to these two departments is about two to four additional TA salaries per year.

One might debate whether the additional TA training costs were actually driven by the change in teaching practices. Many university departments have such TA training programs regardless of the teaching methods in use, as did some departments at UBC before the SEI, such as computer science. However, EOAS and PHAS did not have such programs before the SEI, and it would be difficult to maintain the current teaching without this TA training.

Costs at the University of Colorado. As at UBC, there were no changes in the faculty or administrative costs for teaching in any of the SEI departments. There were only two additional SEI-related costs. The first of these is that MCDB added recitation sections to the large intro course, requiring several additional TAs. This change was made after using SEI funds to fund an experiment that showed these recitation sections had significant benefits for student outcomes. Also, as a result of the SEI, the department became aware that such sections were standard practice for introductory science courses at the university.

The second additional cost at CU came from the addition of undergraduate TAs in a number of courses across the SEI departments. This was done as part of the learning assistant (LA) program,[15] a program whose primary purpose is to recruit science majors to become K-12 teachers. The LA program provides students with early teaching experiences helping in undergraduate science courses and pays them $1,500 per semester while they work as undergraduate TAs. Undergraduate LAs have been integrated into eight of the SEI-transformed courses. Whether they should be considered an added cost is debatable, because supporting undergraduate instruction is only a secondary goal of the LA program, and the majority of the LAs are working in courses that were not part of SEI transformations. In any case, the costs per SEI department for this LA program are small, typically several thousand dollars per year.

In summary, the ongoing instructional costs before and after the SEI transformations are essentially unchanged at both institutions. The primary additional costs have been for TA training.

This analysis includes all the defined costs. There is also the amount of time that the faculty spend on their teaching, which some might argue is a cost that has increased as a result of the SEI, but I do not believe that it is possible to quantify such claims. The job descriptions, standards for hiring and promotion, and the institutional accountability and incentive systems at both these institutions have remained unchanged. So from an institutional perspective, how much time an instructor spends on teaching and how they teach was, and continues to be, entirely a matter of individual preference, with no connection to institutional accountability. Some faculty spend enormous amounts of time on preparing and later modifying highly traditional lectures, while others are using teaching methods introduced by the SEI in an effective way while spending very little time on preparation. From an institutional perspective, none of this is reflected in the instructional cost.

One change that has resulted in a some faculty members spending a little more time on teaching than before the SEI is the addition of homework in courses where previously there had been none. It came as a surprise to me that such courses without homework existed in the sciences, as the educational benefits of required homework are well established. I have since realized that this is one of the areas that differ significantly between disciplines, with physics and computer science having strong traditions of regular graded homework, and biology and earth sciences often only having suggested practice problems with no grading or feedback.

Economics of the SEI Transition Costs

The purpose of the SEI was to carry out a realistic experiment to see if it was possible to achieve widespread change, not to find a model that would minimize the costs of such change.

Table 5.5. Total cost and annual value experienced by UBC

Cost of UBC SEI	$9 million total
Value of credit hours impacted (@ $500 per credit hour)	$70 million per year

For the benefit of other institutions that may consider a similar effort, we have done a simple analysis of the economics of the current model and find it is more attractive than we expected (see Table 5.5). Focusing only on UBC, where the data are cleanest, there are now about 140,000 credit hours a year being taught in a significantly improved form. Because of the UBC funding model, it is difficult to determine the actual cost-per-credit hour, but looking at cost-per-credit-hour data from a number of comparable large public U.S. institutions where it is available, $500 per credit hour is at the low end of the cost range. If we use $500 per credit hour, as of the 2013–2014 academic year the UBC SEI was providing significant enhancement to the value of $70 million worth of credit hours each year. This was accomplished at a total cost of about $9 million as of 2014. Our own measures of improved learning and results from the broader literature on science education research would indicate the transformed courses are likely providing 10–30 percent greater learning. That would equate to an effective increase in the value of those credit hours of between $7 million and $21 million per year. As all current indications are that those improvements in the courses are continuing after the SEI is no longer supporting them, this annual benefit will be continuing for the indefinite future with no additional cost.

Of course, I realize that university budgets are based only on the number of students that enroll, not on the amount of learning that takes place, but this admittedly simple analysis suggests that if one did look at the value of the increases in learning that have been achieved, the SEI model in its current form has been a very a good investment by UBC.

Private Fund-Raising for SEI-Like Activities

There is a reason that the SEI programs were nearly entirely funded from within the universities. There are some unique challenges in raising money from outside the university for such efforts, and these are noted as a warning for anyone considering trying to replicate such an effort.

Despite a large amount of effort in the two years before the SEI began, and for some years after, I had no success in getting funding for an SEI program from existing external grant programs, public or private. Here are some of the likely reasons for this failure.

First, appealing to a donor to support efforts by the university to teach well is perceived to be simply offering the university and its faculty a special bribe to do the job for which they are already getting paid for.

Second, much of the contributions to universities come from satisfied alumni. But satisfied alumni are not going to see changing undergraduate teaching as a high priority, because they were happy with their experience. A dissatisfied alumnus, or a student who withdrew from the university and was not happy with the quality of teaching, is not likely to want to contribute.

Third, most large private donors and foundations have specific priorities that they want to support. The SEIs are quite unusual and do not align well with those priorities. Also, national priorities and attention in education are almost entirely focused on K-12, although there has been a slight shift in recent years.

Fourth, because this effort is quite different from the usual things universities raise money for, the university development offices struggled with how to sell it. Though some potential donors expressed interest to me after initial meetings, the development team often failed to follow up on this interest, probably because of uncertainty as to how to frame a suitable discussion and request.

UBC was eventually successful at raising substantial private gifts to support the SEI, and I am aware of other institutions recently obtaining private donations to carry out SEI like programs. In all of those cases, the SEI was presented with similar arguments used by a high-tech entrepreneur pursuing venture capital: *"There is something new and exciting here, namely, the recent research on learning and its successful demonstration of dramatically improved results in college classrooms. These new, more effective teaching methods are the wave of the future, but they need some start-up funding to get established and into the mainstream. So, modest amounts of one-time funding now can get them over the initial start-up hump and lead to dramatic long-term results."*

SIX

The Post-Mortem: What Worked, What Didn't, and Why

THIS CHAPTER IS an overall analysis of the SEI as an experiment in institutional change. What aspects of the model and implementation were successful and why, and what aspects failed? I focus on aspects of the SEI model for change that seemed good initially, but turned out not to be effective until after adjustments were made. This chapter summarizes the elements that were found to be of greatest importance in improving how the departments taught science. In many cases this involved recognizing and dealing with entrenched barriers in the culture of the departments and the institutions.

The SEI was an experiment in change, designed to answer a vital question: Is it possible to scale up the use of research-based instructional practices and support changes in the teaching culture and practices at a department level? Looking across all the results summarized in Chapter 5, it is clear that it was successful in achieving large-scale change—the teaching of hundreds of courses by hundreds of science faculty was improved, enhancing the instruction of many thousands of student credit hours each year. However, it is also clear that the degree of success was quite variable across departments and institutions. The degree of success in improving teaching methods in any given department is determined by the complex interplay of three basic elements (see Table 6.1).

First are the factors that determine how individual faculty members make decisions about their work in general and teaching specifically. Second are the departmental cultures and how departments function as

Table 6.1. Factors determining SEI impact on department

Faculty decisions	Departmental culture and function	Effectiveness of SEI model at supporting change
External incentives Personal satisfaction Fears of unknown	Leadership and management Distribution of responsibility and authority Course ownership and oversight Curriculum problems Perverse incentives	Elements that clearly worked (competitive grant program and embedded science education specialists) Elements that worked after modification (specific commitments, funding contingent on progress, SES training, focus on willing faculty) Elements that failed (improved efficiency, data driven, sense of urgency)

organizations. Third is how well the SEI model supported change in that context of departments and faculty decisions. It is revealing to dig into the details of those three elements as they are likely to be important to any effort to change the teaching at colleges and universities. This is done below, starting with the SEI model.

Elements of the SEI Model That Clearly Worked

Competitive grant program. A competitive grant program for departments with substantial funds at stake produced widespread attention and discussion of undergraduate education and how they might improve it in every department. Such discussions were quite novel, and in many cases were the first exposure of the faculty in the department to science education research and its findings. In many cases it also mobilized those interested in improving teaching to act with sufficient energy to achieve broad support across the department.

Science education specialists emedded in departments. SESs within the departments proved to be a highly effective way to provide the necessary knowledge, expertise, and time-saving assistance in transforming courses and faculty teaching. It is hard to imagine how the results shown in Chapter 5 could have been achieved without them. I believe that the three critical elements to their success were: they were hired by and seen to work for the departments; they had extensive disciplinary knowledge; and they were well trained in research-based teaching and how to work effectively with faculty members.

Elements of the SEI Model That Worked after Modification

In the initial years at both institutions there were major problems in many areas, particularly in figuring how to establish and maintain good SES-faculty working relationships in departments. In later years the program functioned much more smoothly and effectively due to the recognition of the importance of the items listed below and addressing them suitably.

In terms of overall management and success of the SEI at the central level, the most important characteristics were persistence and flexibility in approaching departments, learning what they need to do to be successful, and providing them with encouragement and pressure to do what is needed. There were four major changes that were found to be necessary in the SEI general management.

Greater oversight of departments and more specific commitments and timelines. First was requiring the department proposals to contain specific commitments in terms of deliverables and timelines. That means giving lists of what courses were to be transformed on what schedule and which specific faculty members were to be involved. In many cases, the timelines turned out to not be realistic, and often things took longer for quite valid reasons. However, requiring this level of detail, including teaching assignments three years in advance, at least laid out specific targets that departments would have already discussed when the proposal was submitted. This avoids the problem that was often encountered in the early days of the SEI, where every faculty member is saying, "Yes, that is what the department was going to do, but I never promised to do anything. Get somebody else."

Making funding contingent on progress. Second, was making funding for a department contingent on progress, rather than on being fully committed upfront. This provided needed accountability and meaningful oversight to ensure that commitments were being met and money was being well spent.

More extensive and formalized initial and ongoing SES training. Third was establishing a formal and extensive training program for SESs, and making sure that both the departments and the SESs understood from the beginning that this would require a substantial investment of time, particularly during their first semester.

Focus on changing willing faculty members rather than particular courses. Although the original vision of the SEI was to systematically transform the curriculum, starting with the introductory courses and working upward, that was unrealistic. It did not properly recognize that the important change is in the faculty, and the courses themselves are secondary. The outcome was more successful when departments focused on finding receptive faculty and providing them with the support and incentives to make changes in whatever courses they were teaching. Of course, priority was given to courses that would affect more students over smaller specialty courses, but what was most important was to build up an ever-increasing number of enthusiastic faculty who would pursue, demonstrate, and champion the SEI teaching goals.

A particularly severe manifestation of the error in focusing on courses involved large introductory courses taught in multiple sections by multiple instructors. Such courses seemed to be the obvious place to start carrying out course transformations. They would greatly benefit from having a high-quality set of materials and assessment items that could be used by multiple instructors; it would save time and create a course that was more effective and consistent for student learning. However, as noted in Chapter 3, it was discovered that multiple-instructor, multiple-section courses where the instructors were used to having substantial autonomy were much more difficult to transform than single-instructor courses. The instructors were often quite resistant to transforming such courses; some were even resistant to agreeing on common topics and exams. In most cases, the fact that the departments and the leadership had now clearly endorsed such changes made little difference due to long established precedent of no supervision. In several cases, a large amount of SES effort was expended on these courses with little success. This held true in multiple courses across several departments. Progress was eventually made in some of these courses after changes were made in the instructional staff.

I did see a large multiple-instructor course that carried out a major transformation without problems. In this case, a single faculty member was clearly seen as being in charge of the course, and provided centralized leadership. This structure would seem to be generally desirable for maintaining the quality and consistency of instruction in multiple-section courses while reducing instructor preparation time. Although this structure is the norm in some departments, in others, strangely, it is not.

I also recognized the need to modify the original SEI course transformation model of progressing sequentially through developing learning goals, improving assessments, and then designing good instructional activities. While this method of backward design does result in a high-quality transformed course, it was a process that only a fraction of the faculty were willing or able to follow, often because of the difficulty they had initially with formulating good learning goals. From the SESs I learned that it was more successful to take a flexible approach, starting with particular instructional issues of interest and concern to faculty members and help them achieve noticeable and rewarding progress. Only then could they progress to other steps in the process. This shifts the emphasis from a results-centered backward design approach to a people-centered incremental steps and "small wins" approach.

Elements of the SEI Model That Failed

Improved efficiency. Although the SEI was successful in changing the teaching practices of many faculty and in many courses, the SEIs made little progress in improving efficiency by achieving departmentally developed and owned courses with good consensus learning goals and shared instructional activities and assessments. The original hope was that by working together to create effective courses designed to be part of a coherent curriculum, faculty time would be minimized while maximizing student learning, as materials, assessments, and learning goals could be passed along, reused, and improved as faculty rotated through teaching different courses.

The benefits of improving efficiency and effectiveness for individual faculty, and departments as a whole, seemed to be too radical a change from the prevailing culture of individual ownership of courses. The idea of sustaining the course structure by passing on materials and assessments was not perceived as worthwhile by the individuals or departments who would need to do the necessary work. There was no structure within the departments that would assign to someone the job of ensuring that materials were archived and passed along so as to allow faculty to use their time most efficiently, and there were no incentives in place for anyone to take responsibility for such activities at a departmental level.

One of the most obvious manifestations of this difficulty was the lack of transfer of transformed courses. After a course was transformed through the SEI, there was rarely a departmental expectation or plan that other faculty would use the course materials. While many course transformations were sustained, this was due largely to decisions made by individual faculty members rather than a result of departmental policy. The exception was that in a few cases departments arranged to have a transformed course co-taught by the transforming instructor and a new instructor who would be taking over the course.

Widespread collection and use of data. As discussed in Chapter 3, the regular collection and use of data on student learning and attitudes outcomes for ongoing improvement has not been embraced by any department. While we saw some individual faculty members collect and use this type of data, it never became departmental policy or had departmental resources supporting such efforts. It is hard to see this ever happening unless it is driven by the institutional accountability and incentive system.

A sense of urgency about educational improvement. A substantial unsolved problem was how to create a sense of urgency in the SEI work, so that it was seen as a priority both by the department and by individual faculty members. A sense of true urgency—that is, the activity needs to be given high priority because change is needed now, and if it does not occur, serious consequences will ensue—is known to be an important ingredient in creating change within organizations.[1] The original intent of the SEI was to create this sense of urgency by providing resources (money and SESs) for a finite period of time, and encouraging departments to take maximum advantage of these resources before they were gone. Faculty and departments almost never viewed the SEI in this way, resulting in slower change and less than optimum use of the SEI funds.

This was likely the result of the formal incentive system at the institutions and their misalignment with the improvement of teaching. Within the institutions the adoption of better teaching methods was unrecognized by the incentive system in promotion or tenure decisions, or in levels of support for departments. The involvement with an SEI course transformation was nearly always seen by faculty to be a voluntary activity, from which they could withdraw without penalty. If the formal incentive system recognized and rewarded SEI-like improvements in instruction, it is likely that one could accomplish these changes faster and with far less money that was required for the SEI. The process would still be greatly enhanced by having SESs guide the faculty in making changes and acquiring teaching expertise.

Department leaders, most notably in UBC EOAS, were sometimes able to create a modest sense of urgency. The EOAS leadership laid out a detailed plan showing when SESs could work with which faculty on which courses over the course of the SEI, and hence what had to be accomplished by specific dates if the work was to be completed before the end of the SEI funding. This plan was regularly reviewed with the faculty. The formal agreements with deliverables and timelines connected with incentives and signed by faculty members embarking on course transformation efforts also served to elevate the priority of the work.

Factors Influencing Faculty Decisions about Teaching

The success of the SEI depended on many factors, but the most essential was how faculty members decided to make changes in their teaching methods. Through many discussions with SESs and some with faculty, I identified factors that entered into those decisions. Many of these factors have been previously noted by others, although I omit "time," which is usually listed in the literature, in favor of factors that determine prioritization. This is because no faculty member has unused time, rather their decisions are always based on how they prioritize the use of the limited time they have. I found that the concerns that discouraged faculty from adopting new teaching methods and working on course transformation were quite consistent across departments, and the extent to which departments dealt with these concerns was largely the determinant in their SEI outcomes.

The formal (dis)incentive system. The dominant barrier to the adoption of better teaching methods at these and other universities is the formal incentive system, which is actually seen as a disincentive to put time and effort into teaching. The universal concern for tenure-track faculty was how adopting new teaching methods would impact their research productivity. Whenever the issue of changing teaching was brought up, it invariably led to the question "How much time will this take?" A longer conversation made it clear that this really meant "How much time will this take away from my research?" This concern was always raised, even at the proposal stage of the SEI, and in the extreme cases the decision of the department was to not submit a proposal, as it was felt that any benefits to improving teaching would be outweighed by the negative impacts on research.

This priority given to research productivity directly reflects the formal incentive system. At UBC and CU, as at all research-intensive institutions, research productivity is carefully measured and rewarded, but teaching effectiveness is not. There is nothing in the formal incentive system that encourages the adoption of better teaching methods by individuals, or in fact even recognizes that there are different teaching methods that might be used. Similarly, the reporting and incentive system is blind to any collective departmental practices that would make education more effective for the students and teaching more efficient for the faculty. It is entirely reasonable that the faculty and departments align their priorities and efforts with the institutional incentive system, which by default means there is a disincentive to spend time on improving teaching or other aspects of undergraduate education. Another clear manifestation of this prioritization was that some junior faculty explicitly chose to put off working with the SEI until after they had tenure. Much of the success of the SEI, both overall and at the departmental level, was dependent on how well the resources of the SEI were used to counter the pressures of the formal incentive system.

The formal incentive system also served as a disincentive to non-tenure-track instructors adopting effective research-based teaching methods, even though their job descriptions did not include research. This was the result of the fact that the only comparative data relevant to teaching collected by the formal system were student course evaluations. These evaluations do not reflect the quality of the teaching methods used nor the amount of learning achieved[2] and are widely perceived as favoring entertaining lectures and penalizing active learning techniques.

Direct individual incentives provided by departments. Over the course of the SEI, I came to realize the importance of direct SEI incentives to faculty members to make changes in their teaching. Initially I encouraged departments to put nearly all of their funding into supporting SESs, with little funding for explicit incentives to individual faculty. This was a mistake, as it failed to recognize the full importance of the formal incentive system as a barrier. More faculty participated in SEI transformation efforts and with greater enthusiasm when departments provided explicit incentives to them. Such incentives took many forms, and were most effective when they were tailored to the specific needs of the faculty member, often in a way that could benefit their research or free up their time. For some this meant support for a graduate research assistant or partial support for a postdoc, while for others a reduction in teaching load for a term or an additional TA was more attractive.

Another value to direct incentives to faculty was that it meant that SEI Central had a meaningful response if a department failed to follow through on its commitments made in their proposal. Without such direct incentives, SEI Central could and did threaten to cut off funding, but this had little meaningful impact on the faculty, typically the ones failing to fulfill commitments, since the loss of funding would not involve any loss to them personally.

Formal agreements with explicit deliverables. For direct incentives to be effective, however, they had to be connected to formal agreements laying out the expectations as to what the faculty member was to complete. I was surprised at first to discover how often otherwise responsible faculty would fail to live up to informal agreements to carry out course transformation activities, but then I came to understand why this was the case. It was a natural extension of the priority that teaching is given in a faculty member's life. While they all recognize that they have to show up for class, any extra effort devoted to teaching activities is routinely relegated to a lower priority than things like completing research proposals or reports, or solving an immediate problem that arises in their research. These early SEI activities were automatically put into this low-priority classification associated with all teaching activities, and so were often supplanted by other activities.

Initially it was very common for faculty to agree ahead of time to work on a course transformation but then back out at the last minute when they started to seriously consider what time and effort would be required. In many cases an SES was told when hired that he or she would be working with a particular faculty member to redesign a particular course, and then find the faculty member was unwilling. Similarly, projects that faculty members were paid to complete during the summer months seldom were done, and in many cases had barely started by the end of the summer.

EOAS showed us the solution to this problem, as noted in their results section. They established a rather formal-looking agreement that laid out the incentives being provided and a detailed list of expectations and deliverables from the faculty member in return. The agreement was then signed by the department chair and the faculty member, usually after the faculty member reviewed it with the SEI departmental director. Although such an agreement had no formal legal status, it carried with it a powerful message. Such an agreement caused a faculty member to think about this SEI work in quite a different way compared to their usual teaching activities, and as a result it was given much higher priority and was usually successfully completed more-or-less on time. Course transformations under the SEI were no longer seen as part of the "business as usual" of teaching and course preparation, but rather were something more urgent and high-priority, with clear incentives attached and corresponding penalties for failure to complete. The agreement also clarified expectations in advance, so faculty members had a much more realistic view of what would be involved and how they would be working with the SES.

Persuasiveness of educational data. I initially, and erroneously, believed that most faculty members would be convinced to change their teaching methods when faced with research data on the effectiveness of different methods. In reality, data, in the form of discipline-based education research results, had a limited impact on their attitudes. In retrospect this is not surprising, since the psychology literature suggests that people will often discount information that creates cognitive dissonance.[3] Accordingly, I found that if the data about teaching effectiveness conflicted with their core beliefs about teaching and learning, the tenets of their discipline about teaching, or their self-image as a good teacher, faculty could always find ways to discount this data, particularly if the data had not been collected in their classroom with their students. This finding has been supported in other studies.[4]

Perhaps the most dramatic example of this came when I was first presenting the idea of the SEI to the UBC physics and astronomy department, as the first step in the proposal process. I gave my standard presentation, in which I discussed ideas of physics education research and offered data from many studies showing the benefits of research-based physics instruction compared to traditional lectures. At end of the talk, there was heated opposition to these findings and the idea it would apply to UBC physics courses, led primarily by several award-winning faculty members who were known for their charismatic lecture performances. After this argument went on for some time, a young woman stood up. She explained very articulately how she had been an undergraduate at UBC in physics and now she was a graduate student—and that everything I said was exactly correct, in her experience. She said she had gone to all those lectures and had been able to do well on all the exams but had never understood the material. Only now that she was a graduate student and having to teach many of these courses was she beginning to actually understand the physics. (I later learned that she was considered one of the top graduate students in the department.) This statement carried far more weight than all the research studies I had presented. It resulted in a great buzz of discussion in the room, some acceptance that maybe there could be some truth to this, and an acknowledgment that the department needed to look into it further.

Another example of where a local example was found to be far more convincing than published research came from the statistics department. Early on, one faculty member who had been stimulated and encouraged by the SEI carried out cognitive interviews with about a dozen students who had received As in his course the previous year. He found to his shock that almost none of them could explain the most fundamental concept that underlay the entire course. After that, he and others in the department were convinced they needed to change their teaching.

While data are seldom convincing, I found it to be true that science faculty will generally pay some attention to research data and give it some thought, even if they were not convinced by it and of the need to change. However, faculty were strongly biased toward data from their particular discipline and were not influenced by data from other fields; frequently they expressed the belief that what works for teaching in one field, such as physics, may not apply to teaching in other fields, such as chemistry or biology. In contrast to the science faculty, the mathematics faculty largely ignored educational research data—perhaps not surprisingly, as their discipline is not an empirical science.

Rewarding personal experiences. The evidence that seemed to have a bigger impact on faculty changing their teaching methods was more personal. When teaching using interactive research-based methods, faculty consistently found teaching more personally rewarding, because their students were much more engaged in learning and showed both greater interest in the topic and more attention to the instructor than the instructor had previously experienced. Also, the level of intellectual interaction with the students was much higher, and so the instructor felt they were contributing much more to students learning the rich complexities of the subject. I believe that making teaching a more personally rewarding experience is the primary reason that the SEI was successful in the face of the barrier provided by the formal incentive system, and it is the primary hope for the sustainability of these teaching methods. The SEI provided encouragement and support for faculty to try out and learn to use these new teaching methods in a reasonably successful way, but the reason they continue to teach this way, and proselytize about it to their colleagues, is because they found it personally more enjoyable.

Observing a session of a transformed course was a powerful influence on faculty decision-making. This seemed so influential that I tried to make it as easy as possible for faculty to observe such courses. As in most universities, at UBC visiting another faculty member's class was highly unusual and considered quite strange. To counter this barrier, I encouraged departments to make a list of model transformed courses and, after getting permission from the instructors to have visitors to their class (which instructors were always happy to grant), distribute the times and locations of such example courses to all the faculty. At UBC, the job of assembling and emailing to the faculty a list of such sample courses that would welcome visitors from across the college was initiated by SEI Central and later adopted as an activity of the dean's office. This implicitly recognized and endorsed the efforts of faculty who had carried out very successful course transformations, as well as making it easy for other faculty to see these methods being used in practice. Second to the impact of actually observing a well taught class in person was hearing an enthusiastic colleague describing the experience, often in informal settings, such as over coffee or lunch.

Fears of the unknown. Faculty had several specific fears when considering adopting new teaching methods. One was "Will this hurt my student evaluations?" I saw that it was helpful for the department chair to explicitly reassure faculty members that their student evaluations would be handled differently so they would not suffer from lowered evaluations. In reality, this fear was quite unfounded as discussed in the "Student Evaluation" section of Chapter 5, and this concern largely evaporated at UBC as the SEI became well established.

Another faculty concern was "How will I cover all the material?" This was best handled by laying out in detail how the various elements in a transformed course worked to make the learning more efficient and thereby maximize the material that could be covered and learned. We trained the SESs to show faculty how a significant amount of material, particularly simpler transfer of information and mathematical derivations, could be moved out of class via pre-class reading or homework, freeing up time. Thinking about using instructional time more efficiently was often a novel but convincing idea to many faculty. Having examples of actual courses that had been transformed without sacrificing substantial amounts of material, especially in combination with hearing from faculty involved in such efforts, was also usually quite convincing.

Another fear was "How will I keep control of the class?" The idea that instructors will lose all control once they let students start talking with each other in class is a fear held by a nontrivial minority of faculty. Having the faculty member observe a well-run active-learning class was the best way to address this concern. It also helped to show them tricks for running a very large interactive class, such as having a bell that is rung to signal to students that they should stop talking and pay attention to the instructor.

A less common concern was "Won't these methods be helping the weaker students at the expense of the top students?" This was most commonly raised with adopting new teaching methods in upper-division courses. As more students became familiar with these teaching methods, however, the views of the students, particularly many of the strongest students, provided the most powerful and articulate arguments in favor of the new methods. Data on learning for the students at the institution also likely helped. We had data from the upper-division physics courses showing how, once the students had experienced the use of clicker questions and peer discussion in such courses, the students were overwhelmingly (four to one) in favor of such methods. It was also helpful to show faculty that before the students had experienced teaching this way in upper-division courses, they had exactly the opposite opinions (four to one against).

Departmental Culture and Function

The basic requirement for success of a course/faculty transformation was the combination of a trained SES, a willing faculty member, and adequate planning. However, the quality of the management of the department determined how likely it was that all three of these would happen at the same time, and how often there were problems in the implementation. Ultimately, once a department was funded, the primary determinant of departmental success was simply the overall quality of the organization and management of the department. None of the problems or solutions in this regard are unique to the SEI or academic departments; they largely reflect good management planning and practices in any organization, and the failures that result when good practices are not followed. However there are a number of elements in the department culture, such as how "ownership" of courses is perceived, oversight of the large introductory courses, and the status of non-tenure-track instructors that impacted the SEI results.

Leadership and management. The primary leadership in departments is provided by the chair. I saw that the chair played a major role in the SEI success. There were a number of examples where the chair changed during the SEI program. In all of the cases where the new chair was not supportive, the SEI work slowed substantially, and in the cases where the new chair was more supportive, progress improved. There were examples, however, where the chair was quite supportive but there were other important elements missing, and in those cases progress was relatively slow. So it is clear that a supportive chair is necessary but not sufficient to ensure SEI success.

The large impact of the chair is somewhat surprising, as generally the chair has relatively little impact on the life of a science department faculty member. I believe that there are several reasons for this special importance in the context of the SEI. First, the chair plays a major part in the management of the SEI program, either directly or in terms of who is appointed as SEI department director. As discussed above, a productive SEI program requires considerable planning and management of multiple resources: funding, faculty, SESs, and teaching assignments. This is challenging for all departments, and how successfully it is carried out depends mostly on how well the chair understands the complex task and ensures that competent people attend to it. Second, the importance of the chair is amplified by the fundamental conflict between the SEI and the formal incentive system, which penalizes faculty for spending time on SEI activities. A chair who is highly supportive of the SEI work, however, can counter the negative message of the formal incentive system through numerous small rewards to faculty members: desirable teaching or committee assignments, space assignments, salary increments, and so forth. Good chairs also emphasize the importance of the SEI work by having it be on the regular agenda of faculty meetings, bringing to the attention of the department particularly notable accomplishments, and seeking other recognition for participants through teaching awards, thereby encouraging faculty to participate. They can also ameliorate the fears about lower student teaching evaluations. Of course, the chair's effectiveness at fulfilling all of these functions depends on how respected he or she is within the department and how good a leader he or she is. A supportive dean was also important, with the most obvious direct impact being in their selection of department chairs who were supportive of the SEI.

Departmental management of SEI efforts. Whenever a department left SEI oversight up to an existing committee, it did not go well, as such committees were fundamentally reactive. A successful SEI transformation effort required a new structure within the department, such as the formation of a new position and/or committee with the charge of bringing about change in undergraduate teaching.

In addition to having the appropriate structure in place, there must be an individual who has the responsibility to oversee all the SEI efforts. The SEI department director handles many of the general management tasks present in any substantial project, such as hiring and supervising the SESs (including making sure they know what they are supposed to be doing and how they should be prioritizing their time), leading the planning efforts, deciding on allocation of resources, reporting to the department and SEI Central on progress, and so forth.

The SEI department director needs to also carry out an essential management task that is unique to the SEI: putting in place the unfamiliar and somewhat delicate collaborative relationship between faculty member and SES. Four problems were encountered most commonly. First was the faculty member treating the SES as simply a TA, doing little besides carrying out routine instructional tasks on the instructor's behalf. Second, some faculty members failed to meet with the SES or provide materials or feedback in a sufficiently timely manner for the SES to do anything useful. Third, it was a problem when the SES tried to be too helpful and ended up creating most of the teaching materials without the involvement of the faculty member; as a consequence, the faculty member never learned how to do it. Finally, faculty members sometimes decided they were too busy or otherwise not interested in being involved with course transformation, and just told the SES to go away.

With multiple SESs working in a department with multiple faculty members to transform courses and teaching, it required considerable planning by the department to make sure all the pieces of faculty member time, teaching assignment, and SES availability and area expertise were aligned. The necessary planning required people in authority with good organization and planning skills. One particular area that was a common source of problems was teaching assignments. Prior to the SEI, few if any departments had a multiyear plan for which faculty would be teaching which course, but I found that to be essential for good SEI progress.

The role and management of the long-term non-tenure-track instructors within the department was also important for the success of the SEI. Such instructors ended up being powerful supporters of SEI efforts in some circumstances and obstacles in others. Across the various departments, the status, roles, and management of non-tenure-track teaching faculty varied dramatically, as did their involvement in and contributions to the SEI efforts. Non-tenure-track instructors who were very involved in SEI activities tended to be instructors who were highly respected and well integrated into the department. They often rotated through teaching a variety of courses. Many of these teaching faculty became leaders and models of the SEI teaching methods and greatly facilitated adoption within the department as a whole.

There were also many examples where this was not the case. A particularly problematic situation was when there were introductory courses or labs always taught by the same long-term non-tenure-track instructors. These cases often (though not always) led to such instructors being quite disconnected from the departments as a whole and essentially unsupervised, and the courses and their goals often were at odds with the thinking of the regular faculty. Often, these instructors taught sections of large multi-section introductory courses, which contributed to the difficulties in transforming these courses. These problems were usually not recognized until the SEI became involved, but then the department often felt incapable of addressing the situation because it would involve too much conflict.

Course ownership and oversight. A fundamental aspect of the culture of departments that was very relevant to the SEI work was their view about course ownership. At one extreme, the courses are owned and defined by the department and the faculty teach the courses that they are assigned in line with expectations set by the department. At the other extreme, the departmental control stops at the course name and number and the faculty member is free to teach whatever and however they want in the course. I found that views about course ownership were variable but tended to be embedded within department cultures, and those views had a substantial impact on the willingness of the faculty to engage in SEI activities. This was most apparent when departments were first considering submitting proposals. In some departments, the sense of individual ownership of any course that a faculty member might teach was so strong that there was overwhelming opposition to the idea of following any sort of guidelines as to best practices, such as those laid out by the SEI. There was also strong opposition to the idea that a faculty member who might be called upon to teach a carefully transformed course would be expected to adopt the learning goals, materials, and methods that were developed.

The stronger the culture of individual ownership of courses in a department, the more difficult it was to get faculty to embrace the SEI transformation model. In some cases, such individual course ownership was felt to be a matter of "academic freedom," although it is difficult to see how one can claim the concept of academic freedom would apply to allowing a faculty member to teach in an ineffective manner or fail to cover the material listed in the official course description.

Even in departments where there was a sense of departmental ownership of some courses, there existed upper-division specialty courses that were always taught, and hence "owned" by a single faculty member. The fraction of courses transformed across the departments is quite dependent on the fraction of the courses offered by a department that are these specialized upper-division courses. A detailed examination of all the courses that have been transformed shows that relatively few upper-division (especially fourth-year) courses are transformed for any SEI department. Nearly all such courses have relatively small enrollments, are often offered only once every few years, and are typically taught by a single faculty member who is an expert in the subject. Very seldom—if ever—does any other faculty member teach the course or have any involvement with it. All these factors tend to make it harder to carry out the transformation of such courses, and for many of the same reasons such courses are a lower priority for transformation within a department when decisions are being made about how best to use SEI resources. If a faculty member teaches only such courses, this also means that it is difficult to impact that person's teaching following the SEI model.

The existence of many upper-division specialty courses can make the fraction of courses changed an unreliable measure of overall impact within a department. Some departments offer, or at least show on their list of courses, a very large number of such upper-level courses—in some cases they account for more than half the courses listed—while other departments have far fewer. I discovered that when there are a very large number of such courses listed, most are seldom taught.

Curriculum problems. The SEI was focused almost entirely on improving how material was taught and learning was assessed. It did not address what material should be taught, other than expecting that learning goals for transformed courses be specific and well articulated. I felt it was the place of the department to decide what should be taught in its courses and curriculum, and that it would not be productive for us to be involved in those decisions. We adhered to that policy, but our deep involvement with the courses and curriculum through working with the SESs did make us aware of the wide variation across departments in terms of how much attention they paid to the quality of their curriculum. In far too many cases, it was apparent that there had been little attention, resulting in problems that negatively impacted student learning. The most serious curriculum problems were tyranny of content, entrenched dysfunctional course design, and poor curriculum alignment.

Tyranny of content. It was not uncommon to have instructors who would agree that these new teaching methods were better but who felt they had to stick with standard lecturing in order to cover all the material traditionally covered in the course. There was often a general consensus that too much material was being covered in such courses for students to learn, but the instructors nevertheless felt compelled to rush through it all, apparently motivated largely by historical precedent and departmental expectations.

Entrenched dysfunctional course design. There were a few other examples of courses in which the selection and organization of topics were firmly entrenched by tradition but made little educational sense. Typically, these courses consisted of a large number of assorted topics established decades ago that were covered rapidly, and which now had little connection to the students' preparation or their future needs. These courses were the only cases where improvements in pedagogy produced no measurable improvements in student learning. Often faculty recognized these courses as problematic, but the department did not have a functional process for fixing them.

Poor curriculum alignment—redundancy and gaps. A less serious but more common problem was poor alignment between courses in a sequence. Our interactions with the SESs, with their deep immersion into many courses within a department and their understanding of student learning, provided insights as to how well the various courses in a sequence supported each other. Due to a lack of clear learning goals, a lack of knowledge by the instructors of the students' prior knowledge coming into their courses, and a lack of oversight by departments as to what the faculty actually taught in their courses and at what level, there were frequently substantial gaps or redundancy in the curriculum as taught. Essential topics of the discipline were omitted, as all the instructors assumed someone else would cover them, and other topics were covered in almost the same form in multiple courses. Also, prerequisite courses, particularly those taught in other departments, did not actually cover the material that the students were assumed to have learned, or covered it in such a manner that it was very difficult for students to see the connections. The presence of multiple-section, multiple-instructor courses (when the instructors taught independently) also contributed to curriculum misalignment, as the different sections of the same course often covered different material. In many cases, each instructor would select the topics he or she liked to cover, in whatever manner he or she chose. All of these factors contributed to poor alignment of the courses, resulting in considerable inefficiency in the use of both student and faculty time.

Fortunately, severe cases of dysfunctional curriculum were relatively rare. Also, in many cases, as faculty adopted SEI methods, they came to better understand student thinking and then recognize problems in the curriculum. Where the department had a suitable process they then took steps to fix the problems. Thus, the SEI has resulted in a number of beneficial changes to the content of courses and curriculum throughout the SEI departments.

Perverse incentives. Although they are not the primary driver of faculty and departmental actions, I encountered some perverse incentives within the system that reward faculty and departments for teaching that drives away students from the major. Science departments generally spend more per credit hour on upper-level courses and labs, and support does not directly track enrollment so there is a financial disincentive to introducing more effective teaching if that results in more students being successful in introductory courses and choosing to enroll in upper-level courses in the field. The increased financial burden on departments of having more upper-division students was raised by several department chairs in the early days of the SEI. As a result, the UBC dean went on record as promising to shift funding to compensate for any increased upper-division enrollments for departments participating in the SEI.

Likely connected with the financial issues, some departments had established grading policies that limited the number of students that receive high and/or passing grades, independent of the amount the students learned. As noted in Chapter 3, this led to some conflicts within departments when improved teaching methods led to notable improvements in student learning and exam performance relative to previous years. In one extreme case, math had an unspoken rule that a large fixed fraction of students in their gateway course for the major should be failed each year. In this case, we made changing that policy a condition for SEI funding.

In summary, the success of any effort to carry out widespread improvement in the quality of teaching will depend on the complex interaction of many factors. I found that, with suitable flexibility and adaptation, the SEI approach was able to address many of the important factors. However, there are many others that are deeply embedded in the culture and functioning of the departments that play an important role.

Coda

THE SCIENCE EDUCATION INITIATIVE SHOWED that it is possible for large research-intensive science departments to make major changes in their teaching. Most faculty adopted innovative research-based methods, and as a result experienced teaching as a far more rewarding activity than they had found it to be using traditional lectures. Their students attend class more and are far more interested in learning the subjects and benefiting from instructors' expertise. Advancing the craft of teaching has become much more of a shared goal and focus of collaborative intellectual activity in these departments, with faculty sharing methods and results and seeking out ideas from others for novel ways to solve instructional challenges. These faculty did find that it takes time to learn to teach in this new way, because there is substantial expertise to be acquired, but that given suitable support, the time investment is not much greater than that required to create a new course. The results are perceived to be well worth the effort.

However, this majority experience did not come about easily or automatically, and was far from universal in all departments. As arguably the largest experiment of its kind, the SEI revealed a great deal about what it takes to bring about widespread educational change in the context of a large research-intensive university.

Here I attempt to distill from all the preceding chapters the most important lessons I have learned from this experiment. This Coda is intended to serve as advice to any deans, department chairs, or faculty members who desire to improve how their institutions teach science (or for that matter, most any subject). If you count yourself among those ranks, know that your ultimate goal must be to convince faculty and departments that teaching well is not merely a function of knowing one's subject and having a suitable personality. It requires expertise based on established principles of learning and the knowledge of research-based practices that apply those principles to teaching in a specific discipline. To bring about such a change in beliefs and associated teaching practices, your three top priorities should be to provide incentives, to support departmental change, and to maximize faculty buy-in.

Provide incentives. First, you need to appreciate how powerfully the formal incentive system undermines the goal of improved teaching. The evaluation and incentive systems used in universities do not recognize that there is research on learning and that there are fundamental differences in the effectiveness of different teaching methods. Faculty universally perceive official incentive systems as penalizing any time taken away from research to innovate or adopt innovations in teaching. Automatically this causes faculty to place a low priority on efforts to improve how they teach. If you are like me and lack the power to change your institution's well-established incentive system, your first priority must be to find ways to counter it with informal incentives. Such incentives need to exist at both the department and the individual faculty member levels. What I have found is that, once a group of faculty has been somehow induced to spend time learning to use these new methods reasonably effectively, the greater personal satisfaction they receive from teaching in this manner proves more than sufficient to keep them teaching in the new way.

The incentives should start with getting deans (and ideally other administrators) to convey the importance of teaching-improvement efforts in both their public communications and private discussions with faculty. Deans can also ensure that department chairs recognize that improving teaching in their departments is an important part of their jobs, and urge chairs to pass along that message to faculty. Being in the good graces of your dean and department chair is not everything, but it is a significant incentive for most faculty.

Most other incentives require money in one form or another. The SEI showed that it does not cost more to teach using these more effective methods, but it does cost money to bring about change. Money can reduce barriers by providing staff support (in the SEI's case, in the form of science education specialists) to minimize the time it takes faculty members to learn new teaching methods and develop new course materials. Money can also, in smaller but still significant amounts, be used to reward faculty for spending that extra time. Some might cherish release time from teaching or some summer salary, while others might want additional budget for a research assistant or piece of lab equipment. Finally, there are simple social incentives. You should continually look for ways to encourage faculty to communicate to their colleagues about teaching, and about how much more rewarding it can be to teach in these new ways.

How much money is required depends on how strongly the existing incentive system and the departmental culture discourages spending time on improving teaching, as well as the scale of the change desired. At most large research-intensive universities, assuming little change in the institutional incentive system, the cost is likely to be in the range of $50,000 to $100,000 per faculty member, spent over a period of five to ten years.

Support departmental change. Your second priority should be to create change at the departmental level. The departments decide what and how to teach, and so they must be the unit of educational change. I found that an effective starting point was a competitive grant program by which departments vie for substantial amounts of money, based on proposals to improve their teaching. The virtue of such a program is that it gets the department as a whole to discuss its overall teaching needs and opportunities in a way that seldom happens otherwise. I also saw that competing for substantial sums of money can produce a level of planning and commitment in departments that would otherwise not be considered worth the effort. It is important to require that proposals have a substantial amount of detail, specifying which courses and which faculty members are to be involved, and including milestones and timelines for what will be accomplished. You may need to work with departments to help them develop such plans, as they may start with little idea as to what such an effort might look like. You will also need to monitor progress after a department is funded to ensure that commitments are met; the long-standing habit of educational improvement's being treated, if left up to individuals, as a low priority is hard to break. While it is important to commit to several (typically five) years of funding to encourage long-term planning and action, the release of subsequent-year funds should be contingent on adequate annual progress.

I found that, once a department had agreed to pursue educational improvement, the success of its effort was largely determined by the quality of its leadership and administration, and that this quality varied greatly across departments. You will need the explicit support of the chair, but it is also necessary to have new structures and responsible people put in place for managing the program within the department. It never worked to have an existing committee—such as a curriculum committee—handle this job of managing SEI change activities. Such committees are designed to operate in a purely reactive mode, not lead change. We had to pay particular attention to how the department handled the three essential administrative tasks: planning and oversight of the collaborations between faculty and SESs; the associated long-term planning of teaching assignments; and the supervision of the SESs.

In my experience, when a department exhibits conspicuous weaknesses in its administration, that problem is deeply rooted in the history and culture of the department. If you encounter a department that has serious and deeply ingrained dysfunction, my advice would be to simply avoid it. Fortunately, it is more likely that you will encounter departments where there are limited administrative weaknesses which can be managed with a little oversight and pushing—particularly if some of that comes from the dean. Finally, make it your mission to learn from your well-run departments what they are doing to make their change efforts successful, and share those practices.

There are many things that departments can do to counter the low priority accorded to teaching improvement. Consider what is signaled, for example when the chair makes it a regular agenda item at faculty meetings to discuss (and celebrate) the progress of efforts to improve teaching. More formally, explicit written agreements can be drawn up with all faculty members who will be involved in transforming courses. Such agreements might spell out the deliverables and timelines expected, and the rewards the faculty member is to receive for the work.

A key component in every successful SEI department were SESs who combined deep expertise in their particular discipline with expertise in teaching and learning in that discipline. SEI's model specifies that such SESs should be hired and supervised by the department and work collaboratively with the faculty to improve teaching. The SESs act as nonthreatening coaches, providing guidance and support to faculty members as they try new things in their courses. With SES guidance, a faculty member is likely to implement research-based teaching methods in an effective manner from the beginning, and have a positive teaching experience in doing so. The SESs also provide expert and time-saving assistance in developing new course materials and assessments.

Finding SES candidates with the necessary disciplinary knowledge was a straightforward task. Often, they were new PhDs. It was not difficult, either, to find candidates with an interest in education, but it was largely impossible to find ones who also possessed the needed expertise in teaching and learning. I found it was necessary to set up a training program for the new SESs in the relevant research and best research-based teaching methods as applied in their discipline. The training also included guidance on how to work effectively with faculty. We had to make it clear to the SESs and their departmental supervisors that, in addition to the time needed for initial training, SESs need to reserve a few hours per week in perpetuity to spend on professional development, keeping up with the relevant research literature, and learning from each other.

Maximize faculty buy-in. In any SEI-type program, the primary goal has to be convincing faculty to adopt new and better methods in their teaching. This means first convincing them that there is expertise in teaching that is worth acquiring. There are many ways to convince faculty to buy in to the program; incentives, of course, play a large part, as does the use of resources (like SESs) to work collaboratively with faculty members to reduce the barriers to change.

I recommend you start by recruiting any willing faculty member to work on making changes in their teaching, and then accommodate them by adopting whatever process of change works best for them. I started out mistakenly thinking it would be best to transform the courses systematically through the curriculum, starting with the lowest level and working up, and in the process, transforming the teaching of the faculty assigned to teach those courses. What I found works best in the real world is to have far greater flexibility, and to focus on transforming the faculty rather than transforming particular courses. Which courses are easiest and most appropriate to transform will likely vary greatly with the local circumstances, and your top priority should be maximizing the number of faculty members in the department solidly on board with new teaching methods.

You should also stay flexible about how courses are transformed. Some faculty will be happy to carry out a complete overhaul of the course by starting with creating a completely new set of learning goals. But, for many others, an incremental approach works better, for both psychological and logistical reasons. Faculty members often have trouble articulating good learning goals. In the SEI, we found they were more comfortable starting by incrementally adding new teaching methods, aided by an SES, to address specific difficulties that they had noted in their classes. Over time, they then became more comfortable with new ways of teaching, they developed a better understanding of student thinking in their courses, and their teaching and learning goals further evolved.

I still believe that it is important to urge faculty members to start a course transformation by deciding on the learning goals for the course, because having complete and detailed learning goals is so helpful for guiding and sustaining the improvement in instruction. You should appreciate, however, how difficult it is for most faculty to produce such a set of goals when they are first asked to do it, and temper your expectations accordingly. A typical initial response is: "I want the students to understand this set of topics [or these chapters in the textbook] . . ." Faculty often find it hard to express what they mean by "understanding" in the operational terms of what students should be able to do. However, if pushed, over time they usually can develop something suitable, particularly if they are regularly thinking about new teaching methods in the course and what benefits these may provide. Similarly, it was challenging to find ways to effectively measure learning in the pre-transformed courses that could then be compared with post-transformation results. I eventually accepted that this was unrealistic in most cases. It simply conflicted too much with existing institutional norms and expectations. We were able to get good assessments of learning in the transformed courses, with the SESs taking the lead.

Like just about anyone faced with trying something new and unfamiliar, faculty members have a number of fears about using new teaching methods. These can interfere with buy-in, if not addressed. Among most common fears: "It will take too much time away from my research." "The students will not like it, and my student course evaluations will go down." "I will lose control of the classroom, and it will be chaos." "I will never be able to cover all the course material I need to get through." Addressing the first fear largely depends on incentives. The best way to address the others is by arranging for faculty members to observe transformed courses being taught, and having them talk with other faculty members who are teaching transformed courses. I found these direct observations and conversations to be more effective at calming such fears than any data. We also developed short handouts for faculty with specific guidance on how to avoid the other concerns raised.

Finally, when implementing a large-scale effort to improve teaching, you need to have flexibility and patience. You are attempting to change traditions that are centuries old. For many faculty members, one or two years of hearing about these ideas and discussing them with their colleagues may be required before they decide to put a toe in the water and try something different. During this gestation period, you need to provide faculty members with repeated educational exposure and potential incentives. Also, remind them that they do not have to do everything at once or be perfect the first time. Even modest changes will result in improved student learning. By their nature, these teaching methods are somewhat self-correcting. The methods allow the faculty member to better understand in real time how their students are thinking, and hence how to make changes to optimize learning and satisfaction.

APPENDIX 1
SEI Course Transformation Guide

Introduction

As part of the SEI efforts, we created a general guide for faculty for carrying out a course transformation, which includes both designing the course and the educational activities that it will provide and the teaching of the transformed course. This presents the general vision for the design of and teaching of such a course, and so we reproduce this guide here to illustrate what will go into a transformed course.

In some respects, such a course transformation is much like doing a science experiment; there are numerous techniques and details that one needs to know, but one has to also understand the concepts and principles behind the design to be successful. This guide is an attempt to put much of this together in one place in a succinct form, to provide a general perspective for the course transformation. This begins with the basic principles of learning through the details of how to implement various instructional methods in the classroom. In Chapters 3 and 4 we provide a description of the process of the transformation as it was typically carried out in the SEIs by the science education specialists working together with faculty.

Results from research on learning provide a useful conceptual framework for thinking about effective teaching and learning. That leads to a set of general principles about what is important for effective instruction. This framework and these principles, particularly as they apply to science and engineering education, are provided in Carl Wieman, "Applying New Research to Improve

Science Education," *Issues in Science and Technology* 29, no. 1 (2012): 25–32.

Very briefly, the essential elements for effective learning are:

- Students must strenuously and explicitly practice the cognitive components of expertise. This includes the unique disciplinary knowledge, the discipline-specific structures by which knowledge is organized and applied, and the ways in which experts monitor their thinking when learning and problem solving.

- Students must receive effective feedback to guide their thinking while carrying out such practice.

- Students must be motivated to do the hard work required for learning.

- Instruction must recognize and build on students' ideas and existing knowledge.

- Instructional activities need to be consistent with the basic mechanisms and limitations of how the brain processes and remembers information.

With this framework in hand, you now need to look at all the components of a course you will be teaching and map these essential instructional elements onto those components in a consistent fashion, in accord with the constraints and opportunities afforded by the context in which the course is situated. Unless there are a lot of resources and prior information available, it is usually more successful to not carry out a total transformation in the first iteration of the course, but rather to develop the design and then incrementally add things over two or three iterations of the course.

Primary components and relevant constraints on course design

- **Learning goals.** Defined in operational terms of what students will be able to do that demonstrates they have achieved all elements of the desired mastery, both cognitive and affective. These goals should guide the design of all other course components.

- **In-class activities.** Some selection of clicker questions and peer instruction, group activities, worksheets, student presentations, lectures, and other activities to help students actively develop their understanding.

- **Homework.** Pre-class reading, problem sets, projects, papers, and other mechanisms for student to further engage with the topics at their own pace.

- **Assessment and feedback, both formal and informal.** In-class clicker question and discussion, via homework, problem solving sessions, exams, surveys, peer review and discussion, instructor-independent measures of expertise such as concept inventories, and other ways for instructors and students to gauge achievement of the learning goals.

- **Constraints and opportunities.** These typically include the available instructional space, incoming state of knowledge of students (what is known and what are the needs for diagnostics), prerequisites or lack thereof, constraints related to preceding and/or following courses in an established sequence, TA support, grading support, instructor time, technology that can be used to support instruction, and so forth.

There is never enough information available to get a course transformation perfect on the first try under any circumstances, and so you should assume that at least one iteration will be required for fine and/or coarse tuning. Typically the first iteration of a course incorporating these principles provides enormously more information about student thinking, background knowledge, and difficulties than was previously known. This provides a guide for substantial further improvement.

A detailed case study of a major transformation of a course (Introduction to Quantum Mechanics) is available in pdf form at the following URL: http://cwsei.ubc.ca/resources/files/Course_transformation_case_study.pdf.

This Course Transformation Guide contains the following elements, all organized in short, easily digested pieces:

- Guiding principles for instruction
- Specific strategies for instructional activities
- Motivation
- Developing mastery
- Practice and feedback
- Creating self-directed learners
- Creating productive views of intelligence and learning
- Memory and retention

- Suggestions for implementing specific instructional practices
- Creating and using effective learning goals
- First day of class
- Better ways to review material in class
- Basic instructor habits to keep students engaged
- Pre-class reading assignments
- Tips for successful clicker use (a more detailed discussion on the effective use of clickers in instruction is given in the SEI booklet "An instructor's guide to the effective use of personal response systems ["clickers"] in teaching"; see http://STEMclickers.colorado.edu for this guide and videos on effective use)
- Student group work in educational settings
- Creating and implementing in-class activities: principles and practical tips
- What not to do
- Assessments that support student learning
- Promoting course alignment: developing a systematic approach to question development

A periodically updated version of this Course Transformation Guide is available at: www.cwsei.ubc.ca/resources/course_transformation.htm.

Guiding Principles for Instruction

Motivation is important for learning and is an essential part of effective teaching[1]

- Show that the subject is interesting, relevant, valuable to learn, worthwhile, fun . . . Remember that most students do not have the benefit of your experience and perspective.

- Convey that subject is challenging but all students can master it with effort, and why it is worth the effort.

- Convey that you care about all students successfully learning the material.

- Avoid scare tactics, such as saying that subject is really difficult, that many students will fail, and so forth. These turn out to be demotivating to many students.[2]

Think of yourself as a "coach of thinking" rather than as a "dispenser of information"

Learning requires intense mental activity with resulting changes in the brain of the learner.[3]

Feedback that is timely and specific is critical for learning

- Timely, frequent, detailed feedback that shows how to improve (formative assessment) should be provided for all students.

- Give marks for what you value (homework, reading, in class participation, quizzes, pretests . . .). For most students, marks define the expectations and what is important in a course.[2]

1. M. R. Lepper and M. Woolverton, "The Wisdom of Practice: Lessons Learned from the Study of Highly Effective Tutors," in *Improving Academic Achievement*, ed. J. Aronson (New York: Academic Press, 2002).

2. SEI student interviews and focus groups at CU and UBC, as well as other studies.

3. John D. Bransford et al., *How People Learn: Brain, Mind, Experience, and School* (Washington, DC: National Academies Press, 2000); S. Ambrose et al., *How Learning Works: Seven Research-Based Principles for Smart Teaching* (Hoboken, NJ: John Wiley and Sons, 2010).

Teach students how to learn

- Explicitly model expert thinking, being careful not to skip steps that are now automatic for you. Convey how to best learn the material and skills; teach students how to study effectively and what is required for conceptual mastery and retention.[3,4] These are fairly readily acquired skills that are seldom if ever taught.

- Know and teach using the best (proven) practices for achieving learning.[3]

Do's and don'ts for the first week

- Explain why you are teaching the way you are teaching, why the course is worthwhile, what your goals and expectations are. The first classes set the tone for the rest of the term.

- Explicitly work to establish a desired class culture.

- Don't threaten or apologize for what or how you will teach.

Find out what all your students are thinking; recognize they think differently than you do

- Connect to and build on their prior knowledge; explicitly examine student preconceptions.[1,3]

- Probe understanding and adjust teaching as appropriate when you find many are not getting it.

Lay out framework, goals, and context for the knowledge and skills you want students to learn

- Teach the organization and application of the knowledge, rather than just the facts. This is the vital element of mastery that students have the most difficulty with.[5]

4. UBC's SEI guidance for students is accessible at www.cwsei.ubc.ca/resources/student_guidance.htm.

5. See notes 1, 2, 3, and 5 above, and many other studies.

Approach teaching as a challenging subject that can be mastered[1,3,4,6]

- The ability to teach effectively is not innate—it can be learned much like a scholarly discipline.

- Understand how people learn and what processes facilitate learning—these are understood.

- Don't be afraid to copy what works. Use teaching practices that have been proven to be effective; they are readily replicated.

6. Ken Bain, *What the Best College Teachers Do* (Cambridge, MA: Harvard University Press, 2004).

Specific Strategies for Instructional Activities

This document gives strategies to achieve the essential elements of effective learning, motivation, practicing to master expertise, feedback, etc. You should apply these strategies to all the course components. Most of this material is summarized from the excellent book by S. Ambrose et al., *How Learning Works: Seven Research-Based Principles for Smart Teaching* (San Francisco: John Wiley and Sons, 2010). It is recommended that you obtain that book, as it provides more detailed discussion.

- Motivation
- Developing mastery
- Practice and feedback
- Creating self-directed learners
- Creating productive views of intelligence and learning
- Memory and retention

Motivation

Student motivation is a key ingredient in a successful course. Two major components of motivation, as identified by Ambrose et al., are:

I. The **expectation**s that students bring to the classroom, and

II. The **value** that students place on the course material and tasks.

Ways to address students' expectations:

1. **Set attainable goals.** Students are best motivated when they feel optimally challenged—when the course and assignments are challenging, but students feel that they can be successful with some effort.

2. **Let students know your expectations.** Communicate your course goals, and how students can achieve those goals. Align instruction and assessment with those course goals—so that students can practice, and see whether they are achieving those goals. This helps to establish realistic expectations. The use of grading rubrics can help make your expectations of student performance on a task very explicit.

3. **Give students feedback.** Without feedback on their performance, students may lose sight of their progress towards a goal. Feedback is most effective when it is timely (that is, without a long time delay), targeted (that is, focused on a specific student performance on a specific task), and constructive (that is, focusing on strengths and future action as well as weaknesses).

4. **Give students a sense of control and self-efficacy.** Self-efficacy is a very important ingredient to student motivation. Provide students with opportunities to feel successful early in the course. Be sure that your grading standards are seen as fair across students and over time. Provide students some flexibility and choice (for example, on assignment topics). Giving feedback on student progress towards well-articulated course and assignment goals can also enhance students' sense of efficacy and control. Also, help students focus on things that they can control (such as study habits), rather than personal characteristics (such as intelligence). *Avoid* threats and framing your course as competition among students, as these are typically demotivating.

Ways to address students' value of the material:

1. **Highlight the relevance of material and tasks.** Students are motivated to engage with material that relates to their personal interests, everyday

lives, and academic or professional paths. Show students how these skills and ideas will relate to future courses and careers. Create assignments that are authentic and relevant; ensure that homework problems can pass the "Why should anyone care about the answer to this problem?" test.

2. **Get students to reflect on what they have learned.** For example, ask students to write a short paragraph on what they learned from a class or an assignment, and how it applies to an interesting or important problem.

3. **Be enthusiastic.** Your own passion and enthusiasm can be a powerful motivator for students.

For more information about how to effectively use motivation in your teaching strategies, see chapter 3 of Ambrose et al., How Learning Works, *and "Motivating Learning," available at www.cwsei.ubc.ca/resources/instructor_guidance.htm.*

Developing Mastery

In order to develop mastery, students must acquire component skills, practice integrating them, and know when to apply what they have learned. They must not only learn "what" but also "how" and "when" to use knowledge and skills.

Ways to help students learn key skills

1. **Get broad perspectives on necessary student skills.** Decompose tasks by asking, "What would students need to know/know how to do in order to achieve this task?" Use your graduate student assistants in this endeavor, as they more recently struggled with this material. Your colleagues are also good sources of information about necessary student skills, as are professionals outside your discipline.

2. **Identify weak/missing skills and help students practice them.** Early assessments (for example, a diagnostic test of expected prior knowledge), as well as thoughtful analysis of student performance on assignments, can help you identify missing skills. Depending on the number of students exhibiting this lack of mastery, you can either devote class time and resources to the issue or provide other resources. Create opportunities for students to work on their mastery of those skills. To address inaccurate prior knowledge (for example, misconceptions), have students make and test predictions, and explicitly address any inconsistencies.

Ways to help students become more proficient

1. **Give students opportunities to practice.** As with other teaching practices, communicate your intent about the practice opportunities, and make your expectations about students' achievement level explicit.

2. **Use productive constraints to reduce cognitive load.** While practicing a skill, it can be helpful to reduce cognitive load by (a) calling students' attention to the key goals and features of a task (so they are not distracted by extraneous features) and (b) simplify tasks to hone in on key skills. Once they become more proficient, the complexity and scope of the task can be increased.

3. **Assess students on their proficiency.** Test students on how well they have integrated the components of complex tasks. This provides alignment between your goals, instruction, and assessment, and gives students feedback on their progress.

Ways to help students learn when to apply their knowledge

There are a wide variety of strategies for helping students learn to transfer ideas to new contexts, which are described in more detail in Ambrose et al., *How Learning Works.* For example:

- **Discuss the contexts** and conditions in which a skill or approach is applicable, and give students practice in doing this. For example, ask them, "Which statistical technique would be used to solve this problem?" or "What questions could this research method be used to investigate?"

- **Ensure that students practice** skills and understanding in many different contexts.

- **Encourage students to generalize** ideas from a specific context to a broader principle.

- **Make use of structured comparisons** to help students identify critical features. For example, you might give two problems that appear different, but use the same underlying principle.

- **Give prompts** to help students make connections between their knowledge and a new problem. For example, "Think back to the bridge we discussed last week."

For more information about how to help students develop mastery, see chapters 1, 2, and 4 of Ambrose et al., How Learning Works.

Practice and Feedback

Practice aimed at achieving specific goals and feedback on progress are critical for learning.

Ways to give students goal-directed practice

1. **Explicitly identify and communicate goals for students.** Make your expectations clear—both for student performance in the course overall and on a given task. These goals can help guide their practice, especially when these goals are stated in terms of what students should be able to *do* at the end of an assignment or a course. Then use rubrics to more specifically define performance criteria for a particular assignment.

2. **Support students in productive practice.** Give students multiple opportunities for practice (readings, quizzes, in-class activities, homework, and so forth) so that they can develop skills and receive feedback. During these assignments, *scaffold* students' development by giving students more support early in learning (for example, by breaking a task into parts for them), and later remove these supports. Create realistic expectations about the amount of practice required by giving guidelines for the amount and type of practice that will be needed. Instead of guessing how long it will take students to do a task, gather data by asking students how long it took them (for example, the last item on a homework set could be "How long did it take you to do this homework?").

3. **Give students positive and negative examples of performance.** What would ideal performance look like? What types of work would *not* meet your goals?

4. **Modify your criteria as your students become more proficient.** Early in the course, determine an appropriate level of challenge by conducting an assessment of student knowledge. As students progress through the course, refine your goals to meet their changing proficiency.

Ways to give students targeted feedback

There are a wide variety of strategies for giving students feedback, which are described in more detail in Ambrose et al. For example:

- **Provide feedback to the class** as a whole about common errors (you can look for common errors in homework or tests, listen in on student discussions during in-class activities and problem-solving sessions, and so forth).

- **Focus your feedback** on key elements of the task, so that students are not overwhelmed.

- **Communicate about strengths as well as weaknesses.** If students have made progress, point that out to them—people are often unaware of the progress they are making.

- **Give frequent feedback,** made possible through use of frequent, smaller tasks.

- **Give real-time feedback.** Collecting group responses through colored cards or clickers lets you give feedback to the whole group.

- **Use student-to-student feedback.** Explicit guidelines can make student comments on each others' work even more valuable.

- **Have students reflect** on the feedback. Require students to incorporate feedback into later work or have them explain what they did wrong. Example from Carl Wieman's teaching: each homework set starts with "Q1. Select a problem from the last homework set that you did incorrectly and explain what you did wrong and what should be done differently to obtain correct answer."

For more information about how to give students opportunities for practice and targeted feedback, see chapter 5 and appendices D and H of Ambrose et al., How Learning Works.

Creating Self-Directed Learners

In order to become self-directed learners, students must learn to assess the demands of the task, evaluate their own knowledge and skills, plan their approach, monitor their progress, and adjust their strategies as needed.

How to help students learn to assess the task

1. **Communicate the nature of the task and check understanding.** Express the goals more explicitly than you might think is necessary, and what students will need to *do* in order to successfully complete the task. Check students' understanding of the task, and give them feedback on their understanding—for example, you might have them express the goal of the assignment in their own words. Be sure to tell students what it is that you do *not* want as well, by showing common student errors in the past.

2. **Give students criteria for success.** Share the criteria that will be used in student evaluation—for example, with a checklist or performance rubric. This helps students generate realistic understanding of the task, as well as learn to monitor their progress towards success.

How to help students evaluate their knowledge

1. **Assess early and often.** Periodic, timely assessments give students opportunity to get practice and feedback so that they can determine where their strengths and weaknesses lie—in time to make corrections before the exam.

2. **Have students assess themselves.** Reduce your grading burden by giving students tasks and have them check their own work using answer keys.

How to help students plan their approach

1. **Provide a plan.** Scaffold students' self-planning approach by providing them your own model for effective planning. This helps them see how a complex assignment might be broken down into pieces or plotted out over time.

2. **Have students create plans; provide feedback on students' plans.** Students might submit their plan as the first part of a complex assignment. This forces them to externalize their thinking, and gives you the opportunity to give them feedback on that plan.

3. **Compare and contrast strategies.** Problems or tasks can be approached in multiple ways; use of different strategies can help students understand the relative merits, particularly if they are given the task of explicitly determining advantages and disadvantages of different approaches.

How to help students learn to monitor their progress

1. **Model metacognition.** Walk students through your own approach to a problem or assignment, identifying different steps and questions that you would ask yourself to check your progress (for example, "Am I making reasonable assumptions?").

2. **Provide strategies for self-correction and reflection.** Students can ask themselves, "Is that a reasonable answer?" "What assumptions am I making?" or "Is this task taking me too long?" Students can also benefit from reviewing classmates' work, especially when given a rubric.

For more information about how to help students become self-directed learners, see chapter 7 and appendices A and C of Ambrose et al., How Learning Works.

Beliefs about Intelligence and Learning

These beliefs have a major impact on student motivation, choice of learning strategies and methods, and the achievement of effective monitoring and self-regulation of learning.

1. **Discuss the nature of learning.** Tell students about the various types of knowledge, from factual recall, to conceptual understanding, to applying those concepts. This can help move them away from an overly rigid view of learning ("you know it or you don't"). Address common misconceptions about learning, to move students away from unproductive ideas (for example, "I'm not a math person"). Discuss the features of learning discussed in this document, such as the impact of practice on performance. Studies by Dweck and others have shown that a student's view of intelligence has a substantial impact on their motivation, approaches to learning, and their academic success. Those who have a view that intelligence is fixed ("There are right-brained people good at math and science and left-brained people who are not") are less successful than those who have a growth mind-set ("Learning and mastery is achieved through hard work rather than innate talent"). These studies have also shown that such beliefs are quite malleable if explicitly addressed.

2. **Encourage students to persevere.** If students have unrealistic expectations about how quickly they will learn something, they may not push themselves when they hit difficulties. Discuss how you or others you know had to work to become expert in a field. Focus students on aspects of their learning over which they have *control,* such as their study habits, rather than external factors such as their level of intelligence or aspects of the course. This helps to increase self-efficacy and a tendency to work through challenges.

3. **Show them the research.** Present research on learning showing how particular types of learner activities and practice are necessary for achieving expertise, and how teaching practices that involve greater student cognitive activity demonstrate greater learning. Show benefits of mentally demanding study strategies (for example, "Test yourself on retrieval and application of ideas," and fully engaged effort to solve hard problems) compared to less effective strategies (for example, reread and review and practice of easy problems, or split-attention study activities).

For more information about how to address students' beliefs about intelligence and learning, see chapter 7 of Ambrose et al., How Learning Works.

Memory and Retention

Introduction: Research on Memory

Memory can be divided into two types: the long-term memory, which has a large information capacity and can remember information for many years, and the working memory, which handles memory and processing of new information over periods of seconds and minutes and has a very limited capacity. Information enters (and leaves) the working memory quickly and easily. It is much harder to get information into long-term memory, and accessing it is also challenging due to interference among the different items in memory during the retrieval process. Repeated retrieval and application of the information, spaced out over time, is the most important element for achieving long-term memory.

The working memory plays a major role in the mental processing that takes place in the classroom, and other similar time-constrained situations, and its limitations have a correspondingly large impact on learning that takes place in that setting. The human working memory has a remarkably small capacity, typically four to seven new items (for example, things not already in long-term memory). The working memory does not just store information, it also carries out basic processing, and so as it is called upon to remember more new items, its ability to process is correspondingly reduced, analogous to a computer with very limited RAM.

The very limited capacity of the working memory has profound implications for the design of suitable classroom activities. It means that ***anything*** that puts additional demands (cognitive load) on the working memory of the student has a cost in what the learner can process and learn. For example, every unfamiliar technical term introduced during a lecture has a significant impact on the capacity of the audience to follow arguments and process the ideas, even if it that term is clearly explained and/or unimportant. Similarly, studies have shown that anything that involves unnecessary input of information or processing during a learning activity has a detrimental effect. Mayer and colleagues have done a series of studies showing how the addition of "seductive details" commonly used by many teachers and textbooks, such as adding amusing anecdotes, attractive pictures, or background graphics that are only peripherally related to the topic, reduce learning.

Strategies to reduce unnecessary demands on the working memory in the classroom

1. Explicitly show how different topics or ideas are linked together, and explicitly show the organization of the class presentation/activities, emphasizing

how the parts are connected. This helps the different topics to be consolidated ("chunked") in the working memory of the students rather than remain distinct, thereby taking up less capacity. Novices often do not recognize these connections that are obvious to experts.

2. Use analogies—this maps complex relationships onto existing relationships already in long-term memory, so the working memory needs only remember the link to relevant part of long-term memory.

3. Use pictures, even simple sketches, to illustrate spatial relationships, rather than relying on verbal descriptions that must be translated into images.

4. Provide worked examples for initial problem solving. Worked examples show the organizational structure and focus the learner's attention on key elements, reducing cognitive load.

5. Use pre-class reading assignments and quizzes to have students review definitions and basic examples before class. See "Preclass Reading Assignments: Why They May Be the Most Important Homework for Your Students," accessible at www.cwsei.ubc.ca/resources/files/Pre-reading_guide_CWSEI.pdf.

6. Keep the use of unfamiliar jargon to an absolute minimum; remembering each new term has a cost.

Strategies for Achieving Long-Term Retention and Useful Access of Learning

1. Provide opportunities and encouragement to students to repeatedly test themselves on retrieving and applying material. The more active the cognitive processing involved in this, the better.

2. Make homework and exams cumulative so that students are reusing and thinking about the ideas multiple times in the presence of new material. Explain why this supports learning.

3. Provide multiple associations ("hooks") between material to be learned and material already in the students' long-term memory.

4. Avoid covering material in a separated sequential fashion, where each topic is covered and tested only once and not revisited. While conducive to a well-organized syllabus, this is not conducive to useful learning. Students need

to build broader associations and to practice sorting out interference between topics when accessing ideas in long-term memory. The additional cognitive processing required to sort out and suppress erroneous interference when studying interleaved topics acts to suppress such interference when accessing information in the future. Too often students will learn and retain that some concept or solution method is associated with chapter 4, covered in week 6, but they will not develop the useful expert-like associations of the material with a suitable range of contexts, concepts, and problem types that will facilitate the desired access from long-term memory.

5. Provide practice activities that explicitly build specific "expert" associations—those commonly recognized and used by experts. Have an assignment that asks students to explain all the ways a new solution method or principle might be used to solve problems associated with topics encountered earlier in the term. Have the students generate general criteria for deciding when this material might be useful.

References on memory and retention:

Michelle D. Miller, "What College Teachers Should Know about Memory: A Perspective from Cognitive Psychology," *College Teaching* 59 (2011): 117–122.

Robert Bjork, "Memory and Metamemory Considerations in the Training of Human Beings," in *Metacognition: Knowing about Knowing*, ed. J. Metcalfe and A. Shimamura, 185–205 (Cambridge, MA: MIT Press, 1994).

R. Mayer et al., "Increased Interestingness of Extraneous Details in a Multimedia Science Presentation Leads to Decreased Learning," *Journal of Experimental Psychology: Applied* 14, no. 4 (2008): 329–339.

R. K. Atkinson et al., "Learning from Examples: Instructional Principles from the Worked Examples Research," *Review of Educational Research* 70, no. 2 (2000): 181–214.

Suggestions for Implementing Specific Instructional Practices

The rest of this transformation guide provides guidance on a variety of instructional practices, both in and out of the classroom:

- Creating and using effective learning goals
- First day of class
- Better ways to review material in class
- Basic instructor habits to keep students engaged
- Pre-class reading assignments
- Tips for successful clicker use
- Student group work in educational settings
- Creating and implementing in-class activities: principles and practical tips
- What *not* to do
- Assessments that support student learning
- Promoting course alignment: developing a systematic approach to question development

Creating and Using Effective Learning Goals

by CU-SEI and CWSEI (2014)

An important first step in course transformation has been to define explicit learning goals for each course which then shape the instruction and assessment. Here we briefly describe the process and benefits of writing learning goals. Learning goals explicitly communicate the key ideas and the level at which students should understand them in terms of what the students should be able to *do*. Learning goals take the form **"At the end of this course, students will be able to . . ."** followed by a specific action verb and a task. For each course, faculty typically define five to ten course-level goals that convey the major learning themes and concepts, as well as topic-level learning goals (also known as "learning outcomes" or "objectives") that are more specific and are aligned with the course-level learning goals. Below are examples of learning goals from an introductory genetics course and a second year physics course. A variety of other examples are available at the SEI learning goals resources link given below.

Examples of Learning Goals from an Introductory Genetics Course (University of Colorado)

Course-level learning goal:

Deduce information about genes, alleles, and gene functions from analysis of genetic crosses and patterns of inheritance.

Topic-level learning goals:

a) Draw a pedigree based on information in a story problem.

b) Distinguish between different modes of inheritance.

c) Calculate the probability that an individual in a pedigree has a particular genotype or phenotype.

d) Design genetic crosses to provide information about genes, alleles, and gene functions.

e) Use statistical analysis to determine how well data from a genetic cross or human pedigree analysis fits theoretical predictions.

Examples of learning goals from a second year physics course (Univ. of British Columbia-UBC)

Course-level learning goal:

Be able to argue that the ideas of quantum physics are true and that it is useful for engineers to know about them.

Topic-level learning goals:

a) Given a simple physical system, be able to draw the relevant potential energy curve needed to model dynamical behaviour.

b) Be able to explain the essential role of the quantization of light as demonstrated by the photoelectric effect in the operation of a photomultiplier tube, a solid state photodetector such as used in motion sensors, and the human eye.

c) Be able to design an experiment for determining the composition of an unknown pure metal based on the photoelectric effect.

d) For an unknown material, be able to analyze whether it is a conductor, insulator, or semiconductor, and then predict what electron energy distribution it must have.

e) Qualitatively design a semiconductor diode that will only allow current to flow in one direction.

The following process of developing learning goals has worked well for course transformations in the SEIs: A working group composed of faculty members who have previously taught a course and those who teach subsequent courses is formed. These working groups typically include a facilitator whose role is to review and synthesize materials, and create learning goal drafts. Learning goals are drafted by referring to materials used by instructors who previously taught the course, with emphasis on homework assignments, exams, and other materials that demonstrate what instructors want students to be able to do. Faculty members who teach subsequent courses communicate what they expect students to know coming into their course. The members of the working group discuss and revise these learning goals until a consensus list is generated, which for any instructor teaching the course would typically cover 70–80 percent of the class time. One of the most critical aspects of writing learning goals is choosing a verb that describes exactly what students should be able to do. Many faculty are tempted to use the verb "understand," but this is not specific—two faculty members could both say "understand" but have completely different expectations as to what students should be able to do. We recommend creating learning goals that convey the relevance and usefulness of any particular content to students. Use everyday language and applications when possible, and minimize the use of technical jargon. Many courses at CU and UBC include goals that focus on skills, habits of mind, and affective outcomes such as: "Students should be able to justify and explain their thinking and/or approach to a problem or physical situation."

Based on our experiences, we formulated a checklist to help instructors create and critique learning goals (below).

Checklist for creating learning goals:

- ❑ Does the learning goal identify what students will be able to do after the topic is covered?

- ❑ Is it clear how you would test achievement of the learning goal?

- ❑ Do chosen verbs have a clear meaning?

- ❑ Is the verb aligned with the level of cognitive understanding expected of students? Could you expect a higher level of understanding?

- ❑ Is the terminology familiar/common? If not, is knowing the terminology a goal?

- ❑ Is it possible to write the goal so it is relevant and useful to students (for example, connected to their everyday life, or does it represent a useful application of the ideas)?

We also aligned the verbs with the cognitive level expected of students. The table that follows shows levels of learning and examples of verbs that match each level, based on Bloom's taxonomy of the cognitive domain.

Levels of cognitive understanding and corresponding verbs

Level	Description	Representative verbs
Factual knowledge	Remember and recall factual information	Define, list, state, label, name
Comprehension	Demonstrate understanding of ideas, concepts	Describe, explain, summarize, interpret, illustrate
Application	Apply comprehension to unfamiliar situations	Apply, demonstrate, use, compute, solve, predict, construct, modify
Analysis	Break down concepts into parts	Compare, contrast, categorize, distinguish, identify, infer
Synthesis	Transform, combine ideas to create something new	Develop, create, propose, formulate, design, invent
Evaluation	Think critically about and defend a position	Judge, appraise, recommend, justify, defend, criticize, evaluate

Benefits

Writing learning goals requires effort and time, but carries multiple benefits. Faculty use learning goals as they plan class time, develop homework, and create exams. All aspects of the course become better aligned, and focus on what faculty most want the students to achieve. Faculty using learning goals report that writing good exam questions becomes easier. At CU and UBC, we have seen that the cognitive level of exams often increases as faculty align the questions with the higher cognitive level of the learning goals.

Sharing the learning goals with students improves faculty-student communication. Learning goals are often posted online and each lecture begins with the relevant learning goals for the day. Surveys reveal that students are overwhelmingly positive about having access to learning goals. The greatest reported benefit is that learning goals let students "know what I need to know," which helps students focus on important ideas and study more effectively.

For departments, writing learning goals has informed, shaped, and aligned the departmental curriculum. By considering the learning goals from multiple courses, departments have discovered that some concepts were taught in an identical manner in multiple courses and other critical concepts were omitted entirely. As a result faculty members who teach different courses have begun to work together so that their goals complement each other and encompass what every student should be able to do by graduation. For instance, some fundamental evolution concepts were added to the CU biology curriculum after this process revealed their absence.

Resources:

www.cwsei.ubc.ca/resources/learn_goals.htm: compilation of learning goals and other resources from the CU and UBC SEIs.

Michelle Smith and Katherine Perkins, "'At the End of My Course, Students Should Be Able to . . .': The Benefits of Creating and Using Effective Learning Goals," *Microbiology Australia,* March 2010, 35–37. http://microbiology.publish.csiro.au/?act=view_file&file_id=MA10035.pdf.

Beth Simon and Jared Taylor, "What Is the Value of Course-Specific Learning Goals?" *Journal of College Science Teaching* 39 (2009): 52–57.

Stephanie Chasteen, Katherine Perkins, Paul Beale, Steven Pollock, and Carl Wieman, "A Thoughtful Approach to Instruction: Course Transformation for the Rest of Us," *Journal of College Science Teaching* 40 (2011): 24–30.

First Day of Class: Recommendations for Instructors

CWSEI, 2014

Set the Environment

The first day of class can have a large influence on students' perception of the entire course. By the end of the first class, you want students to have a good sense of why the course is interesting and worthwhile, what kind of classroom environment you want, how the course will be conducted, why the particular teaching methods are being used, and what the students need to do (generally) to learn material and succeed in the course. It is also important to give the students the sense that you respect them and would like all of them to succeed.

1. Establish Motivation

a. Provide an entry-level preview of the course material and explain why the course material is important and interesting. Avoid jargon as much possible. Where applicable, make connections to:

- Real world/everyday life
- What students know
- What students will need to be successful in future studies or career
- What students are interested in (current events . . .)

2. Personalize the Learning Experience

a. Welcome students to your class—make it clear that you are looking forward to working with them.

b. Introduce yourself, including describing your background and interests in connection to the subject, for example:

- Why you find it interesting and exciting for them to learn
- How it applies to other things you do (research . . .)

(Students—especially those majoring in the subject—say it is inspiring to hear about the instructor's background and research, and how it is relevant to the course.)

c. Introduce teaching team

- If applicable: TAs and anyone else involved that students will be interacting with (could show pictures or have them come to class)

d. Make an effort to find out who the students are and their expectations, motivations, and interests, for example:

- Ask them a series of questions about major, goals, background, etc. (perhaps use clickers or a survey)

- If appropriate, ask them to introduce themselves to other students they will be working with. (Note that this should be used with caution; some students say it makes them uncomfortable if used as a general icebreaker, but it is appropriate to introduce themselves to group members with whom they will be working.)

3. **Establish Expectations** (best if also handed out and/or online, not just spoken)

a. Describe overarching (course-level) learning goals—big-picture view

b. Emphasize that you want them to learn and your role is to support their learning

c. Explain how course will be conducted, what will happen in class, expectations for out of class work, overview of schedule, and marking scheme

d. Explain why you're teaching the way you are teaching, how the different components support their learning. (For examples, see "Framing the Interactive Engagement Classroom," accessible at www.colorado.edu/sei/fac-resources/framing.html.) This is especially important if you are teaching differently than most other courses are taught. For example:

- Teaching methods based on what is known about how people learn

- Students need to play an active part and be intensively engaged in the learning process.

e. Describe (generally) how to succeed in your course

- Learning and improvement take practice and effort; as well as good feedback.

A good activity is to tell students: "1. Think of something you are really good at. Write it down (you don't have to share it with anyone). 2. Now, in one or two words, describe how you got to be good at that thing. 3. On the count of 3, shout out how you got to be good." The overwhelming word shouted will be "PRACTICE." Then talk to them about what kind of practice is the most effective for mastering the material in this course.

- Give general description of how assessments are used for both feedback and marks, leaving details to be read on course website.

- Give advice on how to study.

f. Express that you feel they can succeed if they put in the effort.

4. Details (syllabus, detailed schedule, detailed learning goals, academic conduct, deadlines, rules . . .)

a. Don't go into details during first class; give links to more details on course.

- Could give an assignment involving reading these

5. Other Tips

Good practices	Avoid
Check out classroom before first class (avoid technical problems)	
Start class on time (sends message that you expect them to be on time)	
Telling students you think they can all succeed if they put in the effort (fine to say the course is challenging, as long as also express that it is interesting/worthwhile and doable)	Telling students threatening things, such as you expect some to fail, or lots of students don't like the course and/or have found it very difficult
Address academic conduct in context throughout course (for example, talk about plagiarism when you are giving a writing assignment)	Emphasizing rules and penalties first day (sends message of distrust, and they're not listening anyway)

(continued)

Good practices	Avoid
Provide students with some experiences that give a sense of what future classes will be like	Talking the entire class time
End class on time with slide containing pertinent info (your name, office hours, contact info, website, homework . . .)	Ending class early

In future classes: reinforce these messages periodically in the appropriate context.

Better Ways to Review Material in Class

by Carl Wieman, 2014

A substantial amount of class time is spent reviewing material from previous courses or the previous class meeting. It is very common for instructors to give such review lectures that can occupy one or more classes at the beginning of a term, and/or 5–10 minutes at the start of each class. When we had trained observers at UBC watching the attention of students during classes, it revealed that this form of review was less than useless. Rather than helping students improve their memory and understanding of the material, it primarily diverted their attention to thinking about things other than the class they were in, and this made it harder to get them reengaged when new material was being covered. In retrospect, it is easy to understand why this method of review fails. There is a very well established result from cognitive psychology that familiarity with a topic makes people erroneously believe they understand it. When a person is being lectured on something they believe they already know, they will become quickly bored and start thinking about other things (or checking email, and so forth). This means that students who have previously heard about the topic being reviewed will probably not pay attention, and those students who are not familiar with it will probably quickly get lost in the rapid review.

The better way I found to do review is to replace ALL review lecturing with problems that the students solve in class that cover the material I want to review. This is particularly easy to do if they have clickers. Doing a problem gets them actively thinking about the relevant material and testing their understanding. If they get the problem wrong, and often even if they don't, they are then primed to ask questions and listen to responses and explanations to learn why. Also, if there are things that everyone in the class already knows, I can see that immediately from their problem solutions or clicker responses, and can quickly move on and avoid wasting class time talking about that topic. That leaves more time to spend on the topics where many struggle with the relevant review problem.

A final benefit is that I end up with a good idea of what topics individual students, and the class as a whole, have and have not mastered. As I move on to the subsequent material, I have a vastly better sense of their state of mastery than I previously got from review lectures, and can tailor instruction more effectively.

Another review method: two-stage review

An alternative review format to use at the start of a course is a two-stage review. The two-stage review is patterned after the successful two-stage exams now used in a variety of science courses at UBC. (See Georg W. Rieger and Cynthia E. Heiner, "Examinations That Support Collaborative Learning: The Students' Perspective," *Journal of College Science Teaching* 43, no. 4 [2014]: 41–47, and references therein, accessible at www.cwsei.ubc.ca/SEI_research.) This has similar and possibly greater benefits. Give the students a quiz in class that has the review problems on it, have them do it individually and turn it in, and then have them do a group quiz in groups of three or four and turn in one answer sheet per group. The resulting discussion will provide nearly all the students with the primed and targeted review that they need. The instructor will then only have to worry about dealing with those students whose individual tests indicated they have seriously deficient backgrounds, and dealing with those topics where there are widespread deficiencies. During the group test portion, the instructor should listen in on the various group conversations. That is likely to reveal any widespread difficulties that can then be immediately addressed after the completion of the group test. There would also be a variety of more subtle benefits to this exercise having to do with classroom dynamics, and, as mentioned above, the instructor will know much more about their students' prior knowledge as they move on to subsequent material.

There is a fear that starting the first day with a difficult test will set the wrong tone for the course, so it is best to introduce the two-stage review with a statement like: "This is a carefully designed set of practice problems for your review and discussion, to help you prepare for the upcoming material. This will have no influence on your course grade, except in that they may help you to be better prepared to do well in the course."

A two-stage review was implemented in a UBC science course in the spring of 2014. The third-year course built on topics covered in the second year prerequisite course, but the instructor knew that the students had a variety of backgrounds in that material. Overall, the experience was very positive for the students and instructor, and the instructor learned of some misconceptions that many of the students had.

Basic Instructor Habits to Keep Students Engaged

by Carl Wieman, 2010

It is best to start doing all of these at beginning of the term.

1. **Pay special attention to the back of the room, particularly in a lecture theater.** Walk up aisle as frequently as practical, look at back of room frequently, call on students at back in preference to students in front, repeat student questions so the class can hear, ask students to speak loudly when asking or responding to a question, regularly ask students in back if they can see what is on screen or board and hear what is being said, and don't let chatter in back of the room get out of hand. ALWAYS be conscious of your natural tendency to engage in what effectively becomes a private discussion between you and an individual student in the first or second row.

See end of list for more detailed advice on paying special attention to the back of the room.

2. **When you are talking, regularly stop and ask for questions. Make sure you wait an adequate length of time for response.** What seems like very long time to you is actually short amount of time for a person to collect their thoughts and phrase a question. Instructors typically wait less than two seconds, often less than one, before concluding there are no questions and moving on. A few such very short waits convince students that when you say that you are asking for questions it is just a ritual, and you do not actually want any. Since your time sense in this situation is so skewed, initially you might even use a watch to time yourself to ensure you have waited an adequate amount of time, like twenty to thirty seconds.

3. **If you have a clear impression from facial expressions that students are lost, just say you sense that, and say you need them to ask questions so you can help them, and then wait.** At first they won't believe you, but if you wait long enough (a minute seems like an eternity in that situation) and you look directly at them, someone will ALWAYS ask a question and that starts a discussion. Do that once or twice early in term, and they will learn that you do expect them to raise questions and will then do so quickly.

4. **When a student asks a question, sometimes offer the question to the whole class before answering it yourself.** This reinforces the message that whole class, rather than just you and questioner, should be involved with, and learning from, student questions and answers.

5. **Avoid the tendency to sit back and wait while students discuss a clicker question or in-class activity.** Instead, circulate around the room and listen to them, so you can use what you hear in the follow-up discussion.

6. **After completing a clicker question or in class activity, share student thinking.** If you solicit some answers/explanation or questions from students, rather than you just explaining it, it sends the message that this is about communication and feedback, and it will stimulate ongoing questions from students. If they have written down answers, project some of those (if you have a document projector) or sketch them on the board to share with the class. Sharing answers or calling on a student is not very traumatic for them if they have already worked as group. Call on them to present their group's thinking or answer. Students are normally full of questions after any such activity in which they are obviously engaged, so if you are not getting any questions, you need to figure out what to change.

7. **Define transitions clearly,** such as switching between times for activities involving general student discussion and times when there needs to be general quiet and raising hands before speaking. If you don't, the boundaries get fuzzy, and there can be enough noise in the room that those in back cannot hear and feel left out. Markers that signal a boundary, such as sounding a bell, are quite effective.

8. **Be careful not to send out messages that suppress student engagement.** Obvious examples are suggesting a question is annoying or stupid, asking for questions and only waiting a second, or overlooking raised hands. Some others are:

   a. Jumping in to correct student use of terminology or a small error when main point is correct or relevant. Either ignore the part that is wrong, or correct as an afterthought after discussing the main point.

   b. Suggesting at the outset that a clicker question or activity should be very easy for them. This tends to decrease student motivation to discuss it among themselves or to ask you questions.

   c. Not discouraging highly vocal students who are asking questions primarily to show off rather than to seek an answer. It can send message that asking a question in class is only about showing off.

9. **Avoid facing away from any part of the classroom.** As soon as you are talking with your back to the students, you are conveying that this is a monologue, not a conversation/explanation to them.

10. **Avoid distractions that split their attention.** For example, having a complex image displayed while actually talking about something else. Students will quickly become lost and disengaged.

More detailed advice on paying special attention to the back of the room, particularly in a lecture theater:

a. Walk up aisle as frequently as practical.

b. Very explicitly look at back of room frequently. Call on students at back in preference to students in front, and sometimes explicitly call for answer to question only from students in back. Look at the back and wait patiently for answer when you do so.

c. It is almost impossible not to sometimes overlook raised hands in the back half or sides of even a mid-sized classroom and never realize it. This only has to happen two or three times and you have sent clear message that those students in back are not really part of the class, and they will all stop asking questions from then on. Every now and then apologize for the possibility and encourage students to call out and let you know if this happens.

d. When a student at front says something, if room size allows, ask them to repeat loudly enough and turn so the rest of the class can hear, and regularly remind students when asking questions to do so. In larger rooms (including anywhere you use a microphone), you always need to repeat the student question or comment. Force yourself to do that consistently. Even if it is a room where you will have to repeat question for the back, regularly encourage students to talk as loudly as possible so other students can hear them. The best context for this is when there is a good question—make a comment like "That is an excellent question, everybody in the room should hear and think about that, so can you say as loudly as possible so others can hear?" This sends an explicit message that the whole class is involved and should be learning from student questions, and that it is not just a conversation between you and one student. ALWAYS be conscious of your natural tendency to engage in what effectively becomes a private discussion between you and an individual student in the first or second row.

e. Regularly ask students in back if they can see what is on the screen or board and hear what is being said. Instructors very frequently fail to recognize what cannot be seen or heard from the back. (Whenever you have walked up the aisle, look down to see what viewing is like from student perspective.) Just the act of your checking with them makes them feel more involved and part of the class.

f. A common error in a large classroom is to ignore chatter going on in the back of room and only teach to the front half. DON'T. The earlier in the term you recognize and act on this, the less of a problem it will be. The best preventative measure is regularly walking up the aisle and so you are talking directly to the people in back as much as possible. Also, when you hear chatter in back growing, go up and ask non-talking students in back if they can hear what you were saying and student questions asked from the front. When they say they can't, tell the students to stop talking so other students can hear. (This is a much better tactic than justifying their being quiet on explicit or implicit grounds they are being rude to you.) If that still fails to quiet the chatter, just stop talking and calmly wait while looking at the noisy students in the back.

g. The best preventative to avoid chatter getting out of hand is to early in the term pick someone who seems to be among the worst, find out their name, and then when they start talking, call on them by name, asking them if they have a question. If they are actually talking about class material and do have a question, great. Answer it, then add some comment like, "When you have a question, just raise your hand and ask—we are in the same room, after all." If they were talking about something else entirely and confess to having no question, then gently admonish them to be quiet so students around them can hear the class material. Point out that students often complain about others in back talking in class, making it hard to hear, and they need to be more considerate of their fellow students.

h. When groups are engaged in clicker question discussion or small group activity, try to first walk to the back of class and interact with the students there. Avoid the very common mistake of frequently getting grabbed by students at the front and spending a lot of time with one group and so you seldom get up to the back.

Pre-class Reading Assignments

Why They May Be the Most Important Homework for Your Students

By Cynthia Heiner and Georg Rieger, CWSEI 2012

We usually think of homework as a task, such as a problem set, in which students apply what they have learned in class. But homework can prepare students to learn in future classes. Here we discuss the benefits of pre-reading assignments, report on what students think about pre-reading, and give tips on how best to implement pre-reading assignments to make them effective.

What are pre-reading assignments and what are their benefits?

Traditionally, students are first introduced to a topic in lecture; however, students can read the textbook before coming to class and complete a short quiz on the reading. This is a pre-reading assignment. The first benefit of such assignments is that students will get more out of class if they already know the basic definitions and vocabulary, as well as having already had the chance to work through simple examples and think about concepts at their own pace. This helps control for the variability in background knowledge of the students, and students regularly mention in surveys that pre-reading helps them follow what is covered in class. Also, Louis Deslauriers has monitored the student questions in lectures and noted that student questions are on a cognitively higher level in weeks with pre-reading assignments compared to those in weeks without. Second, by looking at the average responses to pre-reading quiz questions or by directly asking your students what was difficult in the pre-reading assignment, you can gain insight as to which topics your students find difficult. Third, you don't have to spend (much) time on definitions or low-level examples, so you have more class time to focus on the more challenging material.

What students think about pre-reading assignments

Assigning reading is not new. However, in science classes students often do not read the assigned text on a regular basis. So what is different with our pre-reading approach? The assigned readings directly target material used, but not repeated, in upcoming classes and are coupled with targeted quiz questions. This leads students to recognize the textbook as being helpful to their learning.

Typically 85 percent of students report that they read the assigned text every week or nearly every week when the pre-reading assignments are implemented as described here. This has been true across numerous courses spanning several science disciplines. Slightly higher numbers report completing the online quiz (for which self-reports match closely to the computer record). When asked what motivated them to do the pre-readings, the most frequent single answer was the contribution to their grade, but more than half the students said it was because they found the pre-readings "helpful for understanding the material," and "to know what to expect in lectures."

Examples of student comments:

Student A: *"I know that if I complete the pre-reading I will better understand what is going on in the lecture as well as I can figure out where I need to pay the most attention and potentially ask questions."*

Student B: *"I think this forced me to think and was very beneficial to start off the week as I would come into class knowing what to expect and what was expected of me."*

Student C: *"To be honest, I did so because it was for marks. After a while, I didn't mind reading it; and the questions on the pre-reading quizzes help me understand some of the concepts."*

How to implement pre-reading assignments

The pre-reading approach is a variant on "Just-In-Time-Teaching" (JITT), in which every class is preceded by a pre-reading assignment and a quiz with open-ended questions about the difficulties encountered. (See Catherine H. Crouch and Eric Mazur, "Peer Instruction: Ten Years of Experience and Results," *American Journal of Physics* 69, no. 9 [2001]: 970–977.) The instructor reacts to these postings by adjusting the lecture to discuss the difficulties "just in time" for the next class. The full JITT approach requires a strict timetable for the students and the instructor, which is challenging to implement in many courses, particularly ones with large enrollments, and/or multiple sections.

Here we offer a "softer" approach to JITT that provides many of the same benefits. The students get a weekly pre-reading assignment to complete over the weekend, preparing them for the next week of classes. There is a quiz on the reading due before class. There are three key components for the successful implementation of pre-reading assignments: (1) the reading is very specific, (2) the quiz questions explicitly refer to the textbook, and (3) the instructor does not begin class by repeating much of the material in the assigned reading.

Best practices

1. The assignment should focus on what you plan to discuss in class. This creates a clear connection between the reading and the expectations of the students for class.

2. Omit everything that is not necessary. The shorter the assignment is, the more likely the students will actually read it and focus on the key material. Some instructors believe in longer, less focused, readings from which the students are expected to extract the relevant material. This is an unrealistic expectation for a first exposure to the material.

3. The reading should be guided with explicit prompts for the students of what to look for while reading.

4. Give a reading quiz for marks. By assigning marks, you are telling your students that this assignment is important, even if the actual numerical value is small. We have seen that weightings of between 2 percent and 5 percent of the course grade achieve about 85 percent reading completion rates, while assignments without associated marks have much lower completion rates.

5. The questions on the quiz should force the students to read the sections you want them to read and concentrate on the figures that are rich with information. By referring to specific figure numbers (or equations, and so forth) in the textbook, students must at least open the textbook to be able to answer the question.

6. Refer in class to things from the pre-reading—but ***do not*** re-teach them. The point of pre-reading is that the students are expected to come to class prepared with some knowledge. If you re-teach it all, the students will quickly realize that pre-reading is a waste of time and stop doing it. Explain the purpose of pre-reading in your first class and stick with the approach.

7. While there are various quiz options, we have found that a multiple-choice online quiz is better than a paper or clicker-based in-class quiz. In addition to saving precious class time, having the students do the assignment at home with their textbooks open lets them review—before class—their mistakes (and at their own pace). A reading quiz is not a pop quiz—the idea is to prepare students and not to surprise them. Pre-reading assignments should take less than an hour, with the quiz portion, typically around

five questions, taking no more than 10–15 minutes of that time. Use mostly questions that all students could answer with the book, but add in a few that require a little more "reading between the lines." Don't forget: your goal is to draw their attention to something in particular and to motivate, not to trick or overly burden them during their first exposure to the material.

8. It is important that the students understand why and how the pre-reading will be beneficial to them. Explicitly explain your rational and expectations. On the one hand, you expect the students to read the text and try hard to answer the quiz correctly. On the other hand, you do not expect them to "teach themselves" the material nor understand it all completely from the textbook alone. This first exposure gets them started and helps reveal the trouble spots to both the students and the instructor. It is worth repeating the benefits of pre-reading to your students a couple of times during the term.

Tips for Successful Clicker Use

© Dr. Douglas Duncan, University of Colorado, 2008

Including recommendations from members of the Carl Wieman Science Education Initiative. (A longer and more detailed discussion on the effective use of clickers in instruction is given in the SEI booklet "An Instructor's Guide to the Effective Use of Personal Response Systems ('Clickers') in Teaching"; see http://www.cwsei.ubc.ca/resources/clickers.htm for this guide and videos on effective clicker use.)

More than 1,000,000 clickers are in use nationwide, and over 17,000 at CU. Data gathered during the past few years makes it clear which uses of clickers lead to success, and which lead to failure. **Success** means that both the faculty member and students report being satisfied with the results of using clickers.

Clickers have many possible uses: Find out if students have done assigned reading before class; measure what students know before you start to teach them and after you think you've taught them; measure attitudes and opinions, with more honest answers if the topic is personal or embarrassing; get students to confront common misconceptions; facilitate discussion and peer teaching; increase student's retention of what you teach; transform the way you do demonstrations; increase class attendance; improve student attitudes. None of these are magically achieved by the clicker itself. They are achieved—or not achieved—entirely by what *you* do in implementation.

TECHNICAL POINTS:

- Try and get your school to adopt one clicker brand. Students ***hate*** being forced to buy more than one clicker!

- RF (radio) clickers are easier and cheaper than infrared ones.

- Simpler clickers (for example, iClicker) have fewer implementation problems.

- Test your registration system before students do. Deliberately make some mistakes and see what happens. Check early in the semester that all responses are getting credited.

Practices That Lead to Successful Clicker Use

1. Have clear, specific goals for your class, and plan how clicker use could contribute to *your* goals. Do not attempt all the possible uses described above at one time!

2. You MUST MUST MUST explain to students why you are using clickers. If you don't, they often assume your goal is to track them like Big Brother, and force them to come to class. Students highly resent this.

3. Practice *before* using with students. Remember how irritated you get when A/V equipment fails to work. Don't subject students to this.

4. Make clicker use a regular, serious part of your course. If you treat clicker use as unimportant or auxiliary then your students will too.

5. Use a combination of simple and more complex questions. Many users make their questions too simple. The best questions focus on concepts you feel are particularly important and involve challenging ideas with multiple plausible answers that reveal student confusion and generate spirited discussion. Show some prospective questions to a colleague and ask if they meet this criteria.

6. If one of your goals is more student participation, give partial credit, such as 1 point for any answer and 2 for the correct one, for some clicker questions. With some questions it is appropriate to give full credit to all students, such as when multiple answers are valid or when you are gathering student opinions.

7. If your goal is to increase student learning, have students discuss and debate challenging conceptual questions with each other. This technique, *peer instruction*, is a proven method of increasing learning. Have students answer individually first; then discuss with those sitting next to them; then answer again.

8. Stress that genuine learning is not easy and that conceptual questions and conversations with peers can help students find out what they don't really understand and need to think about further, as well as help you pace the class. Students tend to focus on correct answers, not learning. Explain that it is the discussion itself that produces learning and if they "click in" without participating they will probably get a lower grade on exams than the students who are more active in discussion. My students came up with the phrase, "No brain, no gain."

9. Use the time that students are discussing clicker questions to circulate and listen to their reasoning. *This is very valuable and often surprising.* After students vote be sure to discuss wrong answers and why they are wrong, not just why a right answer is correct.

10. Compile a sufficient number of good clicker questions and exchange them with other faculty. The best questions for peer discussion are ones that around 30–70 percent of students can answer correctly before discussion with peers. This maximizes good discussion and learning. There is value in discussion even if a question is difficult and few know the answer initially.

11. If you are a first-time clicker user, start with just one or two questions per class. Increase your use as you become more comfortable.

12. Explain what you will do when a student's clicker doesn't work, or if a student forgets to bring it to class. You can deal with that problem as well as personal problems that cause students to miss class by dropping 5–10 of the lowest clicker scores for each student.

13. Talk directly about cheating. Emphasize that using a clicker for someone else is like taking an exam for someone else and is cause for discipline. Explain what the discipline would be.

14. Watching one class or even part of a class taught by an experienced clicker user is a good way to rapidly improve your clicker use.

Practices That Lead to Failure

1. Fail to explain why you are using clickers.

2. Use them primarily for attendance.

3. Don't have students talk with each other.

4. Use only factual recall questions.

5. Don't make use of the student response information.

6. Fail to discuss what learning means or the depth of participation and learning you expect in your class.

7. Think of clickers as a testing device, rather than a device to inform learning.

If you believe that the teacher, not the students, should be the focus of the classroom experience, it is unlikely that clickers will work well for you.

Be prepared . . . Effective clicker use with peer discussions results in a livelier and more interesting class, for you as well as the students! Expect good results immediately but better results as you become more experienced with clickers. This is the usual experience nationwide.

Student Group Work in Educational Settings

CWSEI and CU-SEI, 2008

Although group work is sometimes hailed as an educational panacea, the realities are considerably more complex. Many studies of group work have been done, and they show a wide variety of results. These range from dramatic improvements in student learning and satisfaction to negative impacts on both. The potential benefits of social interaction on learning are readily apparent. Who has not understood a topic better through explaining it to a colleague and/or having that person raise questions about an explanation? Also, in many situations, peers can provide an effective low cost substitute to individualized instruction by the teacher. However, achieving these and other benefits, such as learning teamwork skills, do not come automatically, and there are clear potential downsides to group work, including the time for organizing groups and dealing with intra-group problems, potential student resentment, more complex grading policies, and difficulties in scheduling and room layout. To achieve the maximum benefit from group work, an instructor must carefully consider the desired educational goals and the benefits, tradeoffs, and pitfalls of introducing different types of collaborative work, and then choose the most suitable type.

Here we briefly review different levels of group work and list the potential benefits and negatives, and what requirements research has shown are needed to ensure a high probability of success.

Levels of collaborative activity—benefits, requirements for success, and negatives

1. Multiple, brief small group discussions in class
(in response to challenging instructor questions or in-class assignments)

A. Benefits: Learn through explanations to others, learn metacognitive skills through analyzing other's reasoning, learn jargon through use in discourse, learn to carry out scientific discourse. Peers provide low level help and feedback, such as catching arithmetic mistakes and avoiding "getting stuck." The stress of speaking in class is reduced, particularly if student is asked what their *group* thought.

B. Requirements: Incorporating this in class is relatively easy—just provide some reason for students to discuss the material with each other. Implementation needs to include some minor reward system or class expectation to promote the group discussion, because otherwise it will not happen spontaneously for many students. Group size should be small (2–4). Two low-effort options for group formation that enhance interaction over just "talk to your neighbor" are: (1) instructor randomly assigns, or (2) students self-organize and register their group online. Such formal groups particularly enhance interaction if students are occasionally required to provide group consensus answers. While it is preferable to have a range of backgrounds and levels in each group, the benefits in this setting are usually not considered large enough to be worth the effort. The benefits are primarily from avoiding groups composed solely of low motivation and low ability students. With mixed groups, the better prepared students can provide explanations to the weaker students, with benefit to both.

C. Negatives: Minor. Time needed to form student groups. Potential disruption due to off-topic discussions in class (usually minor).

D. Other: Opinions vary, but we recommend keeping group composition stable, except where problems.

2. Informal, out-of-class study groups

A. Benefits: Like 1A, plus students can study more effectively by getting low to moderate level feedback from each other. This avoids wasting time from "getting stuck" or overlooking trivial mistakes. Students can succeed at more challenging and complex assignments. Students may find course work more satisfying and enjoyable, and learn teamwork skills.

B. Requirements: Minor. Regularly encourage and discuss the benefits of study groups. Ensure that marking/grading scheme does not appear to penalize collaboration, as discussed below. Provide some form of both group and individual incentives. For example, collaborating can improve grades on assignments, but there are also exams that are closely aligned with assignments. Assignments must be challenging to draw students into meeting for study groups. Make it logistically easy and not socially challenging to form into groups. For large classes, this likely will involve scheduling a room and time for students to meet and/or website for connecting up. Having instructor or TA at these study sessions can draw more students, but it is important that the instructor/TA does *not* provide the answers.

C. Negatives: Negligible. Time needed for elements of B.

3. Formal in-class group activities
(such as tutorials, concept mapping, labs . . .)

A. Benefits: Same as #2, but involves all students. Plus students can develop more teamwork skills.

B. Requirements: Best to have a challenging activity where students work with ideas that are typically difficult to learn and the activity requires them to think about and debate these ideas with each other. Need course structure and space conducive to group work (four per table works well). TAs with role of facilitating group discussion and Socratic teaching works well. Grading options include: only for participation, grading individual work, or grading collective work. Be explicit about why and how collaborative learning is beneficial. If grading collective work, need time and attention devoted to why and how to work in teams effectively, roles and responsibilities of team members, and evaluation of contributions as part of team. Often rotating roles are assigned, manager, recorder, skeptic, and so forth.

C. Negatives: Time and personnel needed to organize facilities and groups.

4. Formal in- or out-of-class collaborative assignments—collective group work and shared marking

A. Benefits: Same as #3, plus reduces time for marking assignments.

B. Requirements: Similar to #3, and a significant goal of the course should be to have students learn to work in teams. Assignments must encourage teamwork, such as being sufficiently difficult or complex that is easier to set up team and work together than to complete as an individual. Assignments that require judgment decisions are found to be most effective at encouraging diverse participation. Groups should be formed by the instructor in a manner that assures equal diversity and skills across groups and is perceived to be scrupulously fair. There must be timely feedback on the functioning of group and a process for dealing with intra-group squabbles.

C. Negatives: (1) There will be some level of student resentment and intragroup disagreements over credit and level of effort. (2) Time required to create groups and deal with logistics. In many courses, groups will not spend the 40 hours of interaction that has been cited as needed to have a highly effective team. (3) Instructors who are not experienced in implementing this can find it difficult to obtain good results.

D. Other: Group size 4–5 is considered optimal, with all visibly underrepresented minority students in a group with at least one other minority student.

5. Learning with fully developed teams

A. Benefits: Same as #4, plus students learn to work as part of team to solve problems and manage projects that would usually be impossible for an individual to complete.

B. Requirements: Major part of course goals needs to be learning teamwork. All of #4B, plus requires more attention to group size, composition, task assignment, general group interaction, and reward system. Majority of course should be team-based project(s). More time and attention devoted to why and how to work in teams effectively, roles and responsibilities of team members, and evaluation of contributions as part of team. Teams should have at least five and preferably six or seven members, and the composition should be as diverse as possible.

C. Negatives: Similar to #4, plus significant time required to create good team-based learning projects.

Group work and marking/grading scheme

If student marks depend on relative student ranking ("grading on curve," "normed," etc.) there is a clear disincentive for a student to collaborate with other students. The inherent contradiction between telling students that they must collaborate, while at the same time penalizing them for helping other students through the marking scheme, will always result in student discomfort and resentment.

References and Resources

C. Crouch and E. Mazur, "Peer Instruction: Ten Years of Experience and Results," *American Journal of Physics* 69 (2001): 970–977. *A good review of peer instruction (falls under Level 1 in this document), including a description of the method and data on effectiveness for improving student learning.*

P. Heller and M. Hollabaugh, "Teaching Problem Solving through Cooperative Grouping. Part 2: Designing Problems and Structuring Groups," *American Journal of Physics* 60 (1992): 637–644. *A good reference on structuring and managing cooperative groups.*

M. Prince and R. Felder, "The Many Faces of Inductive Teaching and Learning," *Journal of College Science Teaching* 36, no. 5 (2007): 14–20. *A nice overview of various forms of inductive teaching that discusses both group and non-group approaches, benefits, and ease (or difficulty) of implementation.*

Team-based Learning: A Transformative Use of Small Groups, ed. Larry K. Michaelsen et al. (2002). *A good reference on team-based learning and also a good reference on group dynamics (chap. 4 by Birmingham and McCord is on research on group dynamics). Also see UBC Faculty of Applied Science website on team-based learning, cis.apsc.ubc.ca/services/team-based-learning.*

Creating and Implementing In-Class Activities: Principles and Practical Tips

CWSEI, 2013

1) Choose a goal or topic to focus the activity

Look closely at your material and ask yourself some of the following questions:

a. What is the most important content or learning goal and how might the activity support that?

b. Are there existing materials (such as a lecture, assignment, or exam question) to base the activity on?

c. Is there an important framework, model, or concept to reinforce?

d. How will it be giving them practice thinking like an expert in the subject?

e. What is most difficult? What gives students trouble? Are there exam questions students do poorly on?

f. Is there a controversy in the material? Is there material that would make a good discussion?

g. What could students work out on their own?

2) Decide how students will engage with the material

The next step is to look at the material you've selected and decide how the students will interact with it. This is key for developing activities. Try to design it so all of the students engage deeply with the content, not just a few.

a. Consider your context. How many students are in your class? How many may require some accommodation? Will you have help administering the activity? How will this work in your particular classroom setting? If the students will work in groups, how large will those be and how will they be formed?

b. What type of activity will be used? If you have difficulty deciding, discuss it with a colleague. Here are a few options that work well with a variety of topics:

i. Think/pair/share (typically 5–15 minutes)—This type of short activity is designed to let everyone engage with the material first individually and then in pairs. First the instructor poses a question, then students spend one minute thinking or writing silently about the idea on their own (you may have to enforce silence, some students will likely try to talk). Then students form groups of two, each partner takes a minute or so describing their thoughts. Finally the instructor facilitates a discussion with the whole class. This activity will usually increase students' responses to questions posed in class.

ii. Worksheets (typically 15–50 minutes)—Write a few questions that lead students through the content in a structured way and photocopy enough for everyone (but see #5d below). Encourage them to work in groups or pairs. The difficulty level should be set so that it is very challenging for most students if working individually, but reasonably doable in groups. An approach that works well is to make the first part relatively easy, so that most groups know how to start, and make later parts more challenging. Adjust the difficulty after running it the first time.

iii. Case study (typically 15–50 minutes)—In a case study, students engage with the content in a real world context. Many people present cases or examples to students in lectures, however it is more effective to give the students material and handouts (for example, graphs, maps, data . . .) that describe the conditions of the case and have them work in groups to make decisions about it. Choose a case that is compelling and requires the students to both analyze the situation and come to a decision or series of decisions and then justify their choices (examples: how to proceed with a project, what to recommend to clients, where to drill, what future changes to expect, how to reduce energy loss, which technique or instrument to use to achieve a goal, and so forth).

3) How will the students be motivated to put in effort?

a. Is it challenging, but doable in groups? Will students see that they are becoming more "expert" at something?

b. Can you connect the activity to a good real world example or something they may do in their future careers?

c. Does it convey why you and others see this topic as interesting and important?

d. Does it involve them making decisions and justifying actions, not simply following set procedures?

e. Does the activity relate to the types of tasks students will be asked to complete on a midterm or final exam?

4) **What product will students generate?**

a. Consider more sophisticated tasks. For example, have students make and justify a **decision** (and perhaps identify the **criteria** used to make a decision), produce a **prediction**, produce a **ranking**, or make a **judgment** (for example, best/worst/most efficient).

b. Consider having students produce a novel representation, such as a specialized graph or sketch.

c. It is usually best to avoid products that depend simply on applying a procedure (such as solving a familiar quantitative problem) or involve extensive writing. These tend to cause more "solo" than "group" work, and are better given as homework. Class time is better spent developing scientific reasoning, and getting feedback.

5) **Logistics and facilitation**

a. Decide how large your groups will be. In a large lecture hall with fixed seats, keep it to 2–3 unless you have them talk with rows in front/behind them. Four in a row doesn't work because the people on the ends get left out.

b. For longer activities, assign roles such as discussion leader, note-taker, or reporter based on arbitrary criteria.

c. Make it very clear what students are expected to do. Ask: "Does everybody know what to do?"

d. Decide how many copies of the activity you will hand out (if you're handing something out). If you have difficulty getting many of your students to work in groups, you can hand out only one sheet per group and make it clear that you expect only one submission per group. On the other hand, it is beneficial for all students to have a copy of their work; some instructors have the students use carbonless copy paper with enough copies for all.

e. During the activity, CIRCULATE and listen to what students are talking about. Look for examples from groups that you could show to the rest of the class for discussion (the doc cam works well in large classes).

f. Plant good questions: if someone asks you a question relevant to everyone, tell them it is a good one and ask them to ask it when you return to the front of the class.

g. Collect something from the students (a completed worksheet, clicker answers . . .) so there is clear accountability for doing the work. You don't need to mark them, but check off for participation and look for useful examples to help you learn more about student thinking and difficulties.

h. Be sure to wrap up the activity effectively. Have a few groups explain their answers. It is more interesting if their answers could be different and spark discussion. Finish by giving your expert summary. Avoid giving a detailed solution that would encourage a student to passively sit through the activity, waiting for you to eventually give them the answers.

6) Assessing the activity

After you've run your activity, reflect on how it went and how it might be improved.

a. Did anything surprise you?

b. Did the students understand what was required? Were they frustrated?

c. Did they engage the way you thought they would? Do you need to adjust the difficulty level?

d. Did they learn what you were trying to teach them (and how can you tell)?

e. Did they enjoy it?

f. Do you need to modify any of your learning goals based on how this went?

7) Other considerations

There are a few other considerations that help in developing activities:

a. Create checkpoints during the activity (for example, a clicker question, or a brief full-class discussion) within longer activities so you can help groups stay roughly in sync.

b. If you know you will have fast groups, add a "bonus" or extra consideration to the end of the activity, one you expect only a few groups will get to.

c. Save class time by having them prepare for the activity. Assign reading and have them answer some relevant questions prior to class.

d. Remember feedback! How are you going to measure and communicate how they've done? Is there a follow-up task that will ensure they think about and use the feedback?

8) Integrating activities into your course structure

a. Aim to make activities a normal, regular part of in-class time.

b. If you're transitioning from dominantly lecture delivery, a good goal is to incorporate at least one 5-minute activity into each 50-minute lecture period, or a longer activity each week. There is probably something in each of your lectures that could be turned into a good activity, particularly if there is student pre-class preparation.

It can be very helpful to bounce your ideas off STLFs (SESs), other faculty, and/or teaching assistants. For more resources, see www.cwsei.ubc.ca/resources/instructor_guidance.htm. Particularly relevant two-pagers on that webpage are "Group Work in Educational Settings" and "What Not to Do: Practices That Should Be Avoided When Implementing Active Learning."

What Not to Do

Practices that should be avoided when implementing active learning

CWSEI, 2013

We and others have written about how to implement active learning in the university classroom, but we have noticed some practices by well-meaning instructors that we feel should be avoided. The numbered items are generally applicable to all types of active learning; there are a few clicker-specific items at the end of the table.

	Don'ts	Comments
1	Don't use active learning without giving students insight into why you are teaching this way	It's important that students feel that the active learning techniques you are using are to their benefit. Some instructors will explain to their students why they are teaching this way (for example, that research shows that people learn much more when they are actively engaged . . .), and others will engage students in discussion about their experience in a particular activity. If you don't address this, students may conclude that you are using less effective techniques or that you are experimenting on them; this can cause resentment and low engagement. It is also good to briefly remind students of the benefits periodically during the term.
2	Don't immediately tell the students the answer and/or explanation	It is usually best to let the students discuss, and then have them share their reasoning with the class.
3	Don't leave activities unresolved	It is important for the students to hear your expert perspective and reasoning. The activity has prepared them to learn from your explanation. Even if you think all the important aspects have come out in the class discussion and/or a large fraction of

(continued)

	Don'ts	Comments
		the students have the correct answer, it is important for you to do a clear and explicit follow-up.
4	Don't forget to make students accountable	Some approaches to building in accountability are: Have the students turn something in (such as a worksheet with all the group members' names on it), use some clicker questions at key points and/or to follow-up on the activity, have random (or all) groups present their results, and so forth. Ensure that clickers are tied to student IDs.
5	Don't have an activity that is not clearly targeting specific learning goals	Activities take time, and therefore should be targeted to important learning goals.
6	Don't overlook motivation	People are much more willing to expend effort if they are intrinsically motivated to do so. It is good to set an activity in a motivating context (for example, a context that is interesting and relevant to the students).
7a	Don't stay in one location of the room during group discussions	By circulating around the room, you can get a better sense of student thinking about the topic (particularly their difficulties and/or misconceptions), and also encourage them to engage in the activity.
7b	Don't spend too much time with one student or group during an activity	Instructors can easily lose track of time when talking with students. This has two detrimental effects: you don't get the benefits of circulating around the room (7a), and many students may become disengaged.
8a	Don't give too many instructions at once and/or make an activity overly complicated	While it is good to make an activity cognitively challenging, introducing too many complications at once adds cognitive load and will confuse and distract students from concentrating on the main goals.

	Don'ts	Comments
8b	Don't make the activity too easy	Trivial clicker questions or activities that have students blindly following steps or repeating memorized facts are a waste of time. Make activities sufficiently challenging so that most students need to discuss and use reasoning to complete them. Consider adding "bonus" questions or problems to keep the high achieving students engaged.
9	Don't expect things to go perfectly the first time you run an activity	If you are running an activity that is new to you, or with a significantly different group of students, it often will not go as planned. Be flexible and modify the activity as needed for the next time. If possible, it is very helpful to test activities in advance with a small group of students and/or discuss it with teaching assistants and other instructors.
10	Don't bite off more than you can chew	Don't try to do more new things in the course than you have time and resources to prepare. You can end up feeling overwhelmed and discouraged. Also, students are usually quite tolerant of an activity that does not go perfectly (#9), but far less tolerant when instructor is obviously disorganized and poorly prepared.
11	Don't forget to clearly indicate the start of an activity	Students will often wait for a signal before starting an activity. Instructors can be expecting the students to start discussing in groups, without realizing the students are waiting for a "Go" signal.
12	Don't lock into a rigid timeline	It's important to be flexible. It is hard to predict the time needed for an activity. Cutting off an activity too soon will leave students frustrated, and going too long will bore students and waste time. Don't use a timer for cutting off clicker responses; instead rely on your judgment.

(continued)

	Don'ts	Comments
13	Don't wait for every student or group to finish	Apply the "75 percent rule" for clicker votes. If 75 percent of the students have clicked in, announce that you will be closing the vote soon (for example, in 10 seconds). For any group activity, you can get a sense of students' progress as you circulate. In longer activities, it is good to have check points where you bring the class into sync.
14	Don't attach high stakes to activities	Accountability is necessary, but assigning a large amount of marks for correctness causes students to seek the "right" answer without worrying about why it is right. Instructors typically give participation points for students who did the activity. If you give marks for correctness, keep this at a low level.
15	Don't embarrass individuals	Be careful in how you react to student statements, particularly if they say something wrong. When calling on individuals, it often is more comfortable for them if you ask them for their *group's* reasoning.
16	Don't get stuck using only one strategy	In order to achieve different types of goals, use a variety of types of activities; if you use clickers, use a variety of question types. Design activities to elicit student reasoning.
17	Don't make comments in advance about the difficulty of activity	Saying things like "I think everyone knows this" or "This should be an easy one" just makes them feel stupid if they don't think it's easy. Also, if you think it is very easy, why use class time on it?
18	Don't rely too much on comments by individual students, or solely on student self-reports about their learning	When there are a few outspoken students, it is very easy to jump to the conclusion that their views are representative of the entire class, but that's often not the case. Use surveys of the entire class or more extensive sampling. Also, student self-reports of what and how they are learning are often inaccurate. Although you should not ignore self-reports, before acting on them you should confirm with other evidence.

	Don'ts	Comments
19	Don't be afraid of a silent moment	Students need time to think after being asked a challenging question.
Clicker-specific don'ts		
	Don't leave out the peer discussion	Using clickers is not good in itself, it is *how* you use them that matters. Peer discussion has been shown to increase student learning, particularly for reasonably challenging conceptual questions.
	Don't show the first vote histogram if you plan to have the students vote twice	In Peer Instruction, students first vote individually and then discuss the question in small groups and vote again. Showing the histogram after the first vote biases the students toward the answer that got the most votes. You can always give a verbal characterization, such as "the vote is split between several options."
	Don't stop the vote collection without warning	Students will rush to put in an answer if they think you might cut off the vote without warning.
	Don't go into "police-mode" for catching students with multiple clickers or not participating enough	Talk with individual students if you see that they are clearly off-task or have multiple clickers (doing the voting for students who are absent), but don't make it a big focus. It needlessly distracts the rest of the class.
	Don't limit yourself to questions with only one right answer	Some of the best peer discussion and whole-class discussions are around questions with more than one defensible answer. For example, you could ask "Which is the best answer?" or "Which is the most efficient method?" In the follow-up discussion, you could ask students what would have to change about the situation to make a particular answer the "best."

Further resources (including materials developed by CWSEI and CU-SEI and links to other useful resources) are available at www.cwsei.ubc.ca/resources.

Assessments That Support Student Learning

CWSEI, updated 2014

Key points and factors from the review paper "Conditions under Which Assessment Supports Student Learning," by G. Gibbs and C. Simpson[1]

Key points (extensive references to data supporting all these points are listed in the original article)

From the students' point of view:

- What is tested in a course dominates what students think is important and what they do.
- Effective feedback is the most powerful single element for achieving learning. Feedback that is not attached to marks can be highly effective.
- Students who focus on picking up cues as to what will be on exams and study accordingly do much better than those who do not. Students often realize this form of studying is not the same as studying to master (i.e., understand and apply) the course material.
- Students prefer courses with a significant marked assignment component, feeling that such courses provide them with more practice and feedback, and the assessment is fairer.

Marked assignments versus exams:

- Much assessment fails to engage students with appropriate types of learning.
- Exam scores correlate very weakly with post graduate performance. Scores on marked assignments are better predictors than exams of long-term learning retention.
- When assignments are a significant fraction of the course mark, the failure rates are 1/3 what they are when the course mark is based solely on exam scores. Students also study and learn in more naïve ways when the mark is based solely on exams. Although not in Ref. 1, there are techniques to minimize cheating on such marked assignments.[2]

Factors that make assessments contribute to learning (and are frequently neglected)

1. Assigned and assessed tasks that:
   - are focused on the most important aspects of the course (tied to learning goals)
   - require extended time to complete
   - are given frequently
   - engage students in appropriate forms of study/effort
2. Students need to have a clear concept of the assigned task and of learning in the discipline. The criteria for setting the mark on the assignment needs to be explicit and understood by the student.
3. The single most important element of assessment supporting learning is the frequency and type of the feedback provided with the assessment. Feedback that supports learning:
   - is frequent and sufficiently timely to the task so that it still matters to the student
   - focuses on student performance and learning, rather than student characteristics
   - is specific and detailed, addresses small chunks of material, and provides guidance for future efforts
   - matches the purpose of the assignment and encourages the student to improve
   - is supported by mechanisms that require the student to attend to and act upon the feedback

Implementing good assessment and feedback without spending excessive time marking

It is particularly challenging to have frequent assignments and timely feedback in large-enrollment classes. Below are a few examples of ways to do this.

- Online, computer graded homework. There are numerous systems for this. (Instructor needs to generate or find source of good multiple-choice questions, many systems provide these.)[3]

- Problem-solving sessions associated with quizzes or homework. This could be informal (groups of students voluntarily get together to work on problems with or without TA or instructor present) or formal (tutorial, recitation, workshop with TA and/or instructor using Socratic approach).

- Peer instruction:[4] during class, pose questions, student discussions about which answer is correct, vote on answer, instructor does short lecture on which answer is correct and why. Works in large lecture halls. (This moves the feedback part into the classroom and shares it between students and instructor. Some coverage of material is moved from lecture to assigned reading.)

- Regular in-class group exercises done in stages that include partial deliverables (sketches, lists, worksheet answers, etc.) which are discussed in class. Simply working in groups provides "instant" peer feedback (as above), and the whole class benefits from feedback that results from the instructor-led discussions at intermediate stages of the exercise.

- Just-in-time teaching:[5] Web-based assignments due a short time before class, followed by discussion/lecture focusing on areas of student difficulty (often involves adjustment of teaching based on responses, for large classes, instructors usually go through a subset of the responses). Can also be implemented as quiz at start of class with electronically collected responses.

- Have some long-answer or essay-type questions on assignments, but only grade some of these (important to be clear to students that they will get some credit on a problem for turning something in, and a subset of those problems will be graded for marks—students won't know in advance which questions will be graded).

- Have multistage assignments with feedback in the middle that students need to use to complete assignment (way to get students to act on feedback).

- Peer assessment (important for instructor to provide good marking rubric). Imperfect feedback from a fellow student provided almost immediately can have much more impact than more perfect feedback from an expert many weeks later. Students learn a lot by *doing* peer assessments—particularly when done as a group activity.[6]

- Self-assessment or reflection assignments (for example, have students grade own work using a rubric created by instructor, or have students go over a problem from previous assignment that they got wrong and explain what they did, and why it was not the correct approach.)

- Two-stage exams:[7] students do the exam individually first, turn their answers in, and then repeat the exam in groups. Students get timely feedback from each other and learn from the exam via reasoning with peers. They usually do significantly better on the group part vs. the individual part.

The bottom line

Teaching students to monitor their own performance should be the ultimate goal of feedback. Continuous support for improving these skills will help students transfer learning to new situations and become effective lifelong learners.

Promoting Course Alignment:

Developing a Systematic Approach to Question Development

By Françoise Bentley and Teresa Foley, 2010
Integrative Physiology Dept. and CU-SEI, University of Colorado–Boulder

When students cannot easily determine the connection between assessments in a course, they often complain that such assignments or activities are "busy work" and "do not help in preparing for the upcoming exam." In order to avoid such discontinuity, it is important that every element of a course be aligned with a set of well-defined learning goals. Using the following systematic approach, faculty can develop a bank of questions that align with a single learning goal. These so-called "suites" of questions can then be used in different settings to measure student learning. For example, one or more questions could be used for formative assessments (for example, a clicker question, quiz, or homework), while a variation of the question(s) could be used on a summative assessment (for example, a final exam). This systematic approach to question development helps faculty focus on their primary educational goals, while it allows students see that the practice they are receiving from assessments is measuring and improving their learning. As an added bonus to using this approach, course exams can be written well in advance of the exam date!

Steps for developing "suites of questions"

1. Start by choosing a learning goal that you would like to assess.

2. Determine the settings where you would like to assess your students (i.e. during lecture, homework, exam, recitation/tutorial, or lab).

3. Develop an initial question for this goal. An application-type question where the students have to predict the outcome of a change in a scenario works best for creating a suite of questions.

   For example, you could create a clicker question that has the students predict the result of increasing a certain variable.

4. Identify what aspects of your question have differing variables/factors that can be changed over a series of questions.

   Using the example above, a related homework question would have students predict the result of decreasing that same variable.

5. Depending on the nature of the question, you can develop at least one exam, one clicker, and one homework question aligned to the same learning goal.

   For example, the corresponding exam question would have students read the scenario and predict if a variable increases, decreases, or causes no change in a particular output quantity.

Example "suite of questions" for a common learning goal

Learning goal: Predict whether a molecule will move across a cell membrane and by what mechanism; explain how concentration and/or electrical gradients influence its movement.

Homework question:

Below is a depiction of a portion of the cell membrane that is positively charged on the intracellular side and negatively charged on the extracellular side. Further in this cell, the concentration of ion X^+ in the intracellular space is high and in the extracellular space is low.

+++++++++++++++++++ intracellular $[X+]_{high}$

<- membrane

- - - - - - - - - - - - - - - - - - extracellular $[X+]_{low}$

Use the figure above to determine what gradients play a role in the movement of ions.

1) Does an **electrical gradient** exist for X^+? If it exists, what is the direction?

a) No. b) Yes, inward. c) Yes, outward.

Clicker question using the same scenario as the homework question:

2) Does a **concentration gradient** exist for X^+? If it exists, what is the direction?

a) No. b) Yes, inward. c) Yes, outward.

In these examples, the homework and clicker questions are assessing the same concept (electrochemical gradients and ion flow), but in multiple ways. For an exam question, you could use a different ion and have the students predict the electrical and concentration gradients of a related scenario.

Exam question:

Consider a typical cell that is temporarily hyperpolarized to –100mV.

What would be the direction of the chemical and electrical forces acting on K^+ while the cell is hyperpolarized?

a) chemical in, electrical in

b) chemical in, electrical out

c) chemical in, no net electrical

d) chemical out, electrical in

e) chemical out, electrical out

f) chemical out, no net electrical

g) no net chemical, electrical in

h) no net chemical, electrical out

i) no net chemical, no net electrical

APPENDIX 2
Guide to interviewing students and faculty

User's guide to interview practices

This document was developed from discussion in CU-SEI SES meetings, September 2007.

Good practices:

General tips for all interviews:

1) Summarize! After every interview try to sit down and write everything you remember about the interview, and any important points you want to remember. Try to do this immediately, or at least within 24 hours of the interview.

    - For more open-ended interviews, try sending your compilation of the interview to the person and ask them if this summary is a correct representation of what they shared in the interview.

2) Phrase questions (in interviews and surveys) so that it is clear to your audience what you are asking.

    - Try out your questions on other people before using them. For concept interviews/surveys, try having faculty members or grad students answer the questions and point out any confusion they had.

3) In most cases, starting out with broad open-ended questions can be helpful before moving on to more specific questions. However, it may result in getting different types of answers than what you expect.

4) Try to finish all interviews by asking if the person you are interviewing has any other comments/questions.

5) Any surveys you give out should be as short as you can make them (while still getting the information you want) to get as many responses as possible. Most people don't like to take the time to fill it out if it's going to take too long (10 minutes is a good length).

Faculty interviews:

1) Instead of sending out mass emails to get faculty members to volunteer for interviews/surveys, try sending emails to individuals or groups of individuals, and address them by name. They may feel more compelled to take the time to volunteer.

2) In the first few interviews, go with very open-ended questions, then use the information from these interviews to come up with more specific questions for later interviews.

3) It may be helpful to bring up what the faculty members' colleagues said about a certain issue to generate more discussion.

    - Might help to get a better "community discussion" going among the faculty in the department.

    - This can be done anonymously, by compiling what all the colleagues said into one lump of information to give to future faculty that you interview.

4) Make sure the faculty members know if the information you obtain from the interview may be shared with others (it is up to you how it may be shared) unless they specify something they don't want shared.

5) When asking faculty about learning goals, try asking them how *they* think the students learn.

6) When interviewing faculty about trying to develop new materials and practices for a course, try to get a better sense of their feelings about learning/teaching by asking them more specific questions.

7) When trying to develop new materials and practices for a certain course, make sure to look at the courses above that course (courses students would move on to take). Find out what the teachers of these higher-level courses expect their students to learn, and what they have found their students to be lacking.

8) Overall good questions to ask:

   - How do they think the students learn?
   - How do they think they should teach to get students to learn?
   - What are students lacking when coming in to their class(es)?
   - What knowledge basis, thinking skills, and affective attitudes would they like the students to have coming into their class, as well as after taking their class?

Student interviews:

1) Similarly to the faculty interviews, start with more open-ended questions, and then find common themes and create more specific questions to ask in later interviews.

2) Start the interview off with some "break the ice" questions to get the student relaxed. Examples:

   - How long have they been here?
   - Their year in school?
   - Did they attend another college before coming here?
   - What is their major? Why did they choose this major?
   - What is their favorite class? Why?
   - Have they had any classes in (your science department) before?
   - What do they want to do after they graduate?

3) Try group interviews/focus groups for finding out about students educational experiences.

- Try emailing the students ahead of time and asking a series of questions, from which you can group them by similar attitudes.
- Information obtained from the students may be richer/more honest; if the students in the group have something in common they may share more.

4) Important aspects to capture about students' educational experiences:
   - How they use their resources (text, course notes, TAs, instructor, etc.).
   - How the information is structured; can they find what they need when studying?
5) When creating concept tests for the students, try sending your questions to faculty members or grad students before giving them to the students you are interviewing to make sure the questions are clear.
6) Give them the survey or concept test and ask them to work through it, talking as they go. Ask them to tell you where they get confused, and if they get quiet, probe them and ask them what they are thinking.
   - Have them explain everything they are doing by drawing pictures, and so forth.
7) When you want to see how students solve problems, try giving them actual problems to solve; rather than asking them how they would solve the problems, watch them do it!
8) Observing or helping at problem solving sessions is very helpful to see where students are having problems understanding the material.
9) To find some common misconceptions, it may be helpful to look on the internet at course materials for elementary students and teachers. They are often full of common misconceptions about science.
10) Another way to get an idea of student misconceptions is to ask the faculty what they think students have misconceptions about.

General tips for all interviews:

1) Try not to make your questions too broad. This may result in getting several different types of answers and it may be difficult to compile the answers and find themes.

2) Make sure your questions are not going to be too difficult for the audience, and also make sure the questions are not going to be confusing or misinterpreted.

3) You don't have to follow your protocol exactly every time you do an interview. If something interesting comes up, pursue it. Follow your gut instincts. Your protocol will evolve as you continue to do interviews and find out what information is helpful.

Student interviews:

1) Try not to finish students' sentences when they are talking through their answers. If they look like they are struggling with their ideas, let them think before helping them.

Let pauses happen. It may seem like a really long pause and an awkward silence, but again, let the students think and give them the time to answer before you interrupt their thoughts with the next question.

APPENDIX 3
Examples of SES advertisements and interview questions

POST-DOCTORAL POSITION IN PHYSICS AND ASTRONOMY EDUCATION

Physics and Astronomy Department University of British Columbia

Applications are invited for two Post-doctoral Fellow positions in the Department of Physics and Astronomy at UBC, to support the department's ongoing program to research and apply innovative teaching techniques.

The successful applicants will work closely with physics faculty to:

- establish sustainable procedures for identifying and effectively using broad-based learning goals, associated assessment tools, and evidence-based teaching techniques in all of our undergraduate courses;
- coordinate a pilot trial of new teaching approaches on a select set of courses; and
- work with others involved in education efforts in the physics and astronomy department and with parallel efforts in other departments in the Faculty of Science.

The initial twelve-month term of employment is normally renewable for a second term, and may be further extended depending on performance and availability of funds. If extended, responsibilities will expand accordingly.

A PhD in a relevant discipline is required, with a sufficient background and experience in Physics and/or Astronomy to be able to teach a range of material at all levels in our undergraduate programs. Applicants must articulate their interest in physics education transformation and research, as well as any prior involvement with such activities (if any).

Applicants should complete the online application form making sure to select "Postdoctoral Position in Physics and Astronomy Education Fac-2014-03."

The positions are available immediately; applications will be reviewed until the position is filled. Three letters of reference should be submitted electronically to jobs@physics.ubc.ca (preferred) or by mail to the address below.

UNIVERSITY OF BRITISH COLUMBIA

DEPARTMENT OF CHEMISTRY

Science Teaching and Learning Fellow—
Carl Wieman Science Education Initiative in Chemistry—2014

1 Position available

The Department of Chemistry at the University of British Columbia invites applications for the position of Science Teaching and Learning Fellow (STLF) for the Carl Wieman Science Education Initiative (CWSEI), a program for the improvement of undergraduate science education at UBC (http://www.cwsei.ubc.ca).

We are currently seeking one individual to work with faculty to develop existing lecture and laboratory chemistry courses. The successful applicants will work with faculty to (a) develop course and program-level learning objectives, assessments, and pedagogy; and (b) develop and implement class materials, including interactive in-class activities as well as pre- and post-tests of learning and attitudes toward chemistry. Publication of research related to the impact of these interventions is expected. For additional information about the STLF positions see http://www.cwsei.ubc.ca/departments/index.html.

Candidates should have a PhD in Chemistry or Chemical Education; excellent organizational, interpersonal, and communication skills; and a strong personal commitment to science education. Familiarity with current pedagogical research at the post-secondary level is desirable. Experience in developing educational materials or curriculum, on-line teaching, and project management will be considered assets.

These appointments will be for one year initially, and may be renewable for an additional year. Appointments will be made at the 12-month lecturer level. The anticipated start date is September 1, 2014.

These positions are subject to final budgetary approval. Salary will be commensurate with qualifications and experience.

UBC hires on the basis of merit and is committed to employment equity. All qualified persons are encouraged to apply. We especially welcome applications from members of visible minority groups, women, Aboriginal persons, persons with disabilities, persons of minority sexual orientations and gender identities, and others with the skills and knowledge to engage productively with diverse communities. Canadians and permanent residents of Canada will be given priority.

Applicants should submit a curriculum vita and a statement of teaching interests and philosophy and arrange to have three reference letters sent directly via e-mail to: STLF2014@chem.ubc.ca.

Deadline for complete applications including letters of reference is August 4, 2014.

Position: Science Teaching Fellow in Physics/PER Postdoc

Applications are invited for a post-doctoral Teaching Fellow in Physics Education in the Department of Physics at the University of Colorado, Boulder. The position is part of the Science Education Initiative (SEI) at CU-Boulder; a program focused on the enhancement of teaching and learning in our undergraduate courses. The successful candidate for the current position will work with the upper-division courses (building on our current efforts in E&M I & QM I, and extending work to E&M II or Mechanics). Candidates must hold a doctoral degree in Physics, possess a strong commitment to science education, have excellent organizational and interpersonal communication skills, and be interested in student learning at the upper-division level. Familiarity with current pedagogy research and assessment techniques or experience in physics education research is not required, but is advantageous.

The Teaching Fellow will serve as the departmental liaison with the Science Education Initiative, directed by Professor Carl Wieman of the Department of Physics. Responsibilities include working in coordination with physics faculty to: develop an integrated plan of course evaluation and innovation; identify specific learning goals that represent faculty-consensus; develop valid assessments of student learning for undergraduate courses; participate in and supervise the development of techniques, materials and practices for improving student learning in the undergraduate courses; and publish assessment tools and findings in Physics education journals. The Fellow will collaborate with and learn from Fellows working towards similar goals in physics and other SEI-funded departments (http://www.colorado.edu/sei), and will collaborate with faculty, post-docs, and graduate students in Colorado's Physics Education Research Group (http://per.colorado.edu).

The appointment is a one-year, renewable appointment with the preference that the successful candidate will be able to commit to the project for two years. The salary will be competitive and commensurate with experience. Applicants should submit a vita and a statement of teaching philosophy and experiences, and have three letters of recommendation sent to:

Job announcement 4/6/06

Science Teaching Fellow

The Department of Molecular, Cellular, and Developmental Biology (MCDB), at the University of Colorado, Boulder, invites applicants for the position of Science Teaching Fellow, to assist with enhancing teaching and learning in our undergraduate courses. Candidates should hold a doctoral degree in Molecular Biology or a related field and have excellent organizational and interpersonal communication skills. However, their primary interest and at least some experience should be in science education. The successful candidate will work both within the department and with other education specialists who are members of the Program for Science Education, directed by Professor Carl Wieman. Specific responsibilities will include working with MCDB faculty who teach our large core undergraduate courses to: specify an integrated set of specific learning goals for these courses, develop strategies for including more active learning, develop and validate assessments of student learning gains, and participate in the development of techniques, materials and practices for improving student learning in these courses.

The appointment is a one-year, renewable appointment with the preference that the successful candidate will be able to commit to the project for two years. Applicants should submit a CV and a statement of teaching philosophy, and have three letters of recommendation sent to:

APPENDIX 4
Sample questions for SES interviews

Departments handle the SES interviews. Typically, the SES is asked to give a talk on their research, or an education-related topic of their choice. They are given information about the SEI and SEI Central. They are asked about their experience working with faculty, how they might handle some common scenarios, their career goals, and their interest or experience with educational research. Special attention is paid to any red flags, such as a personal agenda that doesn't match the SEI goals, or overconfidence in their knowledge about science education.

Sample 1: SES interview

From your previous teaching experience, what is involved in teaching well? In working well with other instructors?

What do you see as your biggest challenge in becoming a STLF? What do you need to learn or be trained on?

Scenarios

1. You are working with a faculty member, Dr. X, to improve an upper-level laboratory course. One of the first things you decide to do is to create learning objectives for the course. How would you manage an email from the faculty member that states their frustration because they "do not see the value creating learning objectives in the lab"?

2. You are working with a faculty member, Dr. X, to develop interactive in class activities for a lecture course. A student from the class emails Dr. X complaining that the interactive activities are a waste of time and the student is paying to hear from the expert in the class, not to talk to his peers. Dr. X is concerned by the email and questions the entire project and the use of in-class activities. How would you respond to the faculty member? How would you advise Dr. X to respond to the student?

3. You are assigned to work on developing materials for a laboratory course. The faculty member teaching the course is not incredibly interested in making changes. How would you connect with the faculty member?

4. You are working with a faculty member who is very excited to work with you on developing interactive activities in class. The faculty member gives you course content and material to build activities around at the very last minute. You are scrambling to complete the activity before class. This repeats. How would you handle this?

Sample 2: SES interview questions, EOAS

Interview questions for STLF candidates, July 2014, EOAS

Morning Meeting

1. What interests you most about this position?

2. What do you think is your most relevant background for the position?

3. *Questions from the committee regarding what the candidate wrote about co-teaching, particularly in assessing the effectiveness of co-teaching on the instructors and transfer of professional skills and pedagogies.*

4. How do you envision your role in supporting faculty in the co-teaching model?

5. How would you help a faculty member who asks for advice on "improving engagement" in his/her course? How would you tell if your advice worked?

6. Describe a teaching/learning situation in which you found a person challenging to work with. Explain how you handled that situation, and the long-term outcome.

7. Describe an assessment tool you have developed (or used). How confident were you that the assessment was measuring what you intended, and on what did you base your confidence?

8. In this job, you will have opportunities to lead and contribute to efforts in research and publication. Are there particular areas of teaching and learning research that are most interesting to you? Do you see ways in which those overlap with the job as you perceive it?

9. How would you like to develop professionally in this position? What skills would you like to develop as an STLF? What do you see as the biggest challenge for you?

10. You'll meet with us again at the end of the day, but do you have any questions for us now?

Afternoon Meeting

1. From your meetings with EOAS faculty members, were there things you learned today about the job, or about expectations of different faculty members, that were new to you? Can we help clarify anything?

2. From your meetings with STLFs from this and/or other departments, do you have questions about what would be expected of you in EOAS? Are there issues the STLFs brought up that we could help clarify?

3. Recognizing that, currently, this is a finite (three-year) project, what do you envision doing afterward?

4. Any other questions about what you have seen, heard, or discussed today?

Sample 3: CU physics SES interview

Areas to probe:

- Faculty interactions—how much experience working with faculty
- Assessing student thinking:
    - Student interviews—what were the goals of your interviews and how have you structured these interviews . . . what works well, what hasn't worked?
    - What are some student difficulties you've seen in teaching E&M?
    - How do you know if your students understand?
- Why are you interested in this position? How does this fit into your career plans?
- Do you have any questions for us?
- Familiarity with literature: What have been some of the PER findings that you find that influence your approach to teaching most?
- Content knowledge: When was the last time that you thought about upper-division E&M content?
- What aspects of your current research are applicable to upper division?

Sample 4: Jackie

Any questions about what you have seen, heard, or discussed today?

Why did you apply for this position?

Core projects

What projects are you most interested in?

What are some ways you could involve more faculty in the type of work that the CWSEI does? (Beyond those involved in the core projects.)

Describe a time when you felt you were not being supervised effectively. That is, something your supervisor was doing was not allowing you to reach your full potential.

Tell me about a time when you felt you needed to speak out and go against what a colleague was saying or doing.

Describe a situation when you successfully "read" someone and were able to guide your actions as a result of figuring out what they needed.

Give an example of a difficult person that you had to deal with. Explain how you handled the situation.

Why do you think many faculty are resistant to active learning strategies?

Describe an assessment tool you have developed. What was it designed to measure, how did you develop the assessment, and how did you know it was working?

What skills would you like to develop as an STLF?

Do you have any questions for me?

What are your salary expectations?

Sample 4: Faculty rating 1

Candidate Evaluation Sheet—Science Teaching and Learning Fellow

The following offers a method for department faculty and others to provide evaluations of job candidates.

Candidate's Name: Your Name :______________

Please indicate which of the following are true for you (check all that apply):

- ☐ Read applicant's CV
- ☐ Read applicant's teaching statement
- ☐ Read applicant's letters of recommendation
- ☐ Attended candidate's job talk
- ☐ Met with candidate
- ☐ Attended lunch or dinner with candidate
- ☐ Other (please explain):

The successful candidate will (1) support faculty members in incorporating evidence-based teaching practices into their courses, (2) assess student learning and faculty professional development, (3) contribute to research in teaching and learning, and (4) teach one course per year in EOAS. Please comment on your impression of the candidate's abilities in these areas (you do not need to address all four).

____

____

____

____

Please rate the candidate on each of the following criteria.

Criteria	Excellent	Good	Neutral	Unable to judge	Fair	Poor
Potential for (evidence of) impact/innovation in teaching and learning						
Potential for (evidence of) educational leadership						
Potential for (evidence of) effective collaboration						
Potential for (evidence of) contributions to research in teaching and learning						
Fit with STLF position						
Ability to make positive contribution to department's climate						
Potential (demonstrated ability) to teach and supervise undergraduates						
Potential (demonstrated ability) to be a conscientious university community member						

Other comments:

Sample 5: Faculty rating 2

*******CONFIDENTIAL*******

Department of Chemistry Science Teaching and Learning Fellow Search

Thank you for taking the time to contribute to this search. On behalf of the Search Committee, your contribution to the process is much appreciated.

It is important that your comments include your name. Anonymous comments will not be considered. Your name will be removed before comments are shared with the search committee.

Please send your comments to Jane Smith in Chemistry (jsudy@chem.ubc.ca) by Feb. 11 at midnight.

This form is meant to be a guide, feel free to elaborate your comments. If you require more space, please use a separate piece of paper/email and attach them to this form.

Your name:____________

Please indicate your status:

Faculty Staff STLF Student Other (please list affiliation)____________

CANDIDATE's name: ______________________________

Please indicate the activity that you participated in and the event(s) that you attended:

I read the following documents from the candidate's application (check all that apply)

- ❑ Cover letter and CV
- ❑ Teaching statement
- ❑ Letters of reference

I attended the candidate's seminar

- ❑ Yes
- ❑ No

I met with the candidate (underline one-on-one or group interview)

- ❑ Yes
- ❑ No

I joined the candidate for lunch or dinner

- ❑ Yes
- ❑ No

Other: ....

CANDIDATE's name: ______________________________

Please comment on the candidate's knowledge and skill in the area of teaching and learning. You may want to consider the candidate's experience in developing teaching/learning resources and assessments, knowledge of education research methods (for example, interviewing, focus groups, surveys), teaching philosophy, and interest in students.

Please comment on the candidate's potential to interact effectively with faculty and students. You may want to consider the candidate's enthusiasm about teaching and learning, personality, work ethic, and leadership ability.

Please provide any other written comments you think would help the committee with its decision.

NOTES

1. The Vision

1. American Association for the Advancement of Science, "Vision and Change in Undergraduate Biology Education: A Call to Action," 2011; Association of American Universities, "Five-Year Initiative for Improving Undergraduate STEM Education: Discussion Draft," Washington, DC, 2011; President's Council of Advisors on Science and Technology, "Engage to Excel: Producing One Million Additional College Graduates with Degrees in Science, Technology, Engineering, and Mathematics," Washington, DC, Executive Office of the President, 2012.
2. Susan R. Singer, Natalie R. Nielsen, and Heidi A. Schweingruber, eds., *Discipline-Based Education Research: Understanding and Improving Learning in Undergraduate Science and Engineering* (Washington, DC: National Academies Press, 2012).
3. Wendy K. Adams, Katherine K. Perkins, Noah S. Podolefsky, Michael Dubson, Noah D. Finkelstein, and Carl E. Wieman, "A New Instrument for Measuring Student Beliefs about Physics and Learning Physics: The Colorado Learning Attitudes about Science Survey," *Physical Review Special Topics: Physics Education Research* 2, no. 1 (2006): 010101.
4. For an extensive discussion of these topics, see James J. Duderstadt, *A University for the 21st Century* (Ann Arbor: University of Michigan Press, 2009).
5. J. Bransford et al., *How People Learn* (Washington, DC: National Academy Press, 2002); Singer et al., *Discipline-Based Education Research*.

6. See, for example, Scott Freeman, Sarah L. Eddy, Miles McDonough, Michelle K. Smith, Nnadozie Okoroafor, Hannah Jordt, and Mary Pat Wenderoth, "Active Learning Increases Student Performance in Science, Engineering, and Mathematics," *Proceedings of the National Academy of Sciences* 111, no. 23 (2014): 8410–8415.
7. P. Ross, "The Expert Mind," *Scientific American,* August 2006, 64; K. A. Ericsson et al., *The Cambridge Handbook of Expertise and Expert Performance* (Cambridge: Cambridge University Press, 2006).
8. E. Redish, *Teaching Physics with the Physics Suite* (New York: Wiley, 2003); Adams et al., "New Instrument"; K. K. Perkins, W. K. Adams, N. D. Finkelstein, S. J. Pollock, and C. E. Wieman, "Correlating Student Beliefs with Student Learning Using the Colorado Learning Attitudes about Science Survey," in *2004 Physics Education Research Conference,* ed. Jeffrey Marx, Paula Heron, and Scott Franklin (Melville, NY: American Institute of Physics, 2005).
9. More details can be found in Carl Wieman, "A New Model for Post-Secondary Education: The Optimized University," Campus 2020, British Columbia Ministry of Advanced Education, http://cwsei.ubc.ca/resources/files/BC_Campus2020_Wieman_think_piece.pdf.
10. Bob Uttl, Carmela A. White, Daniela Wong Gonzalez, "Meta-analysis of Faculty's Teaching Effectiveness: Student Evaluation of Teaching Ratings and Student Learning Are Not Related," *Studies in Educational Evaluation* (in press), doi.org/10.1016/j.stueduc.2016.08.007.
11. Carl Wieman, "A Better Way to Evaluate Undergraduate Teaching," *Change: The Magazine of Higher Learning* 47, no. 1 (2015): 6–15.
12. Wieman, "A Better Way."

2. The SEI Model for Achieving Change

1. Everett M. Rogers, *Diffusion of Innovations,* 5th ed. (New York: Free Press, 2003).
2. Michael Fullan, *The New Meaning of Educational Change,* 3rd ed. (New York: Teachers College Press, 2001); E. Seymour, "Tracking the Process of Change in U.S. Undergraduate Education in Science, Mathematics, Engineering, and Technology," *Science Education* 86 (2002): 79–105.
3. Rogers, *Diffusion of Innovations,* chap. 5.
4. John P. Kotter, *Leading Change* (Boston: Harvard Business Review Press, 1996).
5. Carl Wieman and Ashley Welsh, "The Connection between Teaching Methods and Attribution Errors," *Educational Psychology Review* 28 (2016): 645–648.

6. At CU, the funding arrangement was relatively complex, as it was designed to minimize the SEI's impact on the CU budget. In part, it involved the use of the author's salary money, unspent endowed chair funds that the author had accumulated, and some flexible funds in the central administration and campus budgets.

3. The Process of Making Change

1. The most recent call for proposals, conducted in 2010, is available online at www.colorado.edu/sei/about/funding.htm.
2. See, for example, the EOAS long-term plan, accessible online at www.eoas.ubc.ca/research/cwsei/courses.html.
3. More information about learning goals and various references are available through the website of the Carl Wieman Science Education Initiative. See the resource page at www.cwsei.ubc.ca/resources/learn_goals.htm.
4. Stephanie V. Chasteen, Steven J. Pollock, Rachel E. Pepper, and Katherine K. Perkins, "Transforming the Junior Level: Outcomes from Instruction and Research in E&M," *Physical Review Special Topics: Physics Education Research* 8, no. 2 (2012): 020107.
5. For more on peer instruction, see Eric Mazur, *Peer Instruction: A User's Manual* (Upper Saddle River, NJ: Prentice Hall, 1997); for more on think-pair-share, see Frank Lyman, "The Responsive Classroom Discussion: The Inclusion of All Students," in *Mainstreaming Digest*, ed. A. S. Anderson, 109–113 (College Park: University of Maryland, College of Education, 1981).
6. Michelle K. Smith, Francis H. M. Jones, Sarah L. Gilbert, and Carl E. Wieman, "The Classroom Observation Protocol for Undergraduate STEM (COPUS): A New Instrument to Characterize University STEM Classroom Practices," *CBE-Life Sciences Education* 12, no. 4 (2013): 618–627; additional resources can be found at www.cwsei.ubc.ca/resources/COPUS.htm.
7. Stephanie V. Chasteen, Bethany Wilcox, Marcos D. Caballero, Katherine K. Perkins, Steven J. Pollock, and Carl E. Wieman, "Educational Transformation in Upper-Division Physics: The Science Education Initiative Model, Outcomes, and Lessons Learned," *Physical Review Special Topics: Physical Education Research* 11 (2015): 020110.
8. See www.cwsei.ubc.ca/resources/surveys.htm for descriptions and links to learning attitudes about science surveys.
9. Smith et al., "Classroom Observation Protocol."
10. Carl Wieman and Sarah Gilbert, "The Teaching Practices Inventory: A New Tool for Characterizing College and University Teaching in Mathematics and Science," *CBE-Life Sciences Education* 13, no. 3 (2014): 552–569;

additional resources can be found at www.cwsei.ubc.ca/resources/TeachingPracticesInventory.htm.

4. Science Education Specialists: Agents of Change

1. For more on learning goals, see www.cwsei.ubc.ca/resources/learn_goals.htm.
2. Rachel E. Pepper, Stephanie V. Chasteen, Steven J. Pollock, and Katherine K. Perkins, "Facilitating Faculty Conversations: Development of Consensus Learning Goals," Paper presented at the Physics Education Research Conference 2011, Omaha, Nebraska, August 3–4, 2011, http://www.compadre.org/Repository/document/ServeFile.cfm?ID=11870&DocID=2718 (accessed December 22, 2016); Stephanie V. Chasteen, Katherine K. Perkins, Paul D. Beale, Steven J. Pollock, and Carl E. Wieman, "A Thoughtful Approach to Instruction: Course Transformation for the Rest of Us," *Journal of College Science Teaching* 40, no. 4 (2011): 24–30.
3. See www.cwsei.ubc.ca/resources/surveys.htm for examples.
4. For more on student engagement, see Erin S. Lane and Sara E. Harris, "A New Tool for Measuring Student Behavioral Engagement in Large University Classes," *Journal of College Science Teaching* 44, no. 6 (2015): 83–91. It describes the Behavioral Engagement Related to Instruction (BERI) protocol developed by Lane and Harris. For more on instructor practice, see Michelle K. Smith, Francis H. M. Jones, Sarah L. Gilbert, and Carl E. Wieman, "The Classroom Observation Protocol for Undergraduate STEM (COPUS): A New Instrument to Characterize University STEM Classroom Practices," *CBE-Life Sciences Education* 12, no. 4 (2013): 618–627.
5. Wendy K. Adams and Carl E. Wieman, "Development and Validation of Instruments to Measure Learning of Expert-like Thinking," *International Journal of Science Education* 33, no. 9 (2011): 1289–1312.
6. Stephanie V. Chasteen, Rachel E. Pepper, Marcos D. Caballero, Steven J. Pollock, and Katherine K. Perkins, "Colorado Upper-Division Electrostatics Diagnostic: A Conceptual Assessment for the Junior Level," *Physical Review Special Topics: Physics Education Research* 8, no. 2 (2012): 020108; Bethany R. Wilcox and Steven J. Pollock, "Validation and Analysis of the Coupled Multiple Response Colorado Upper-Division Electrostatics Diagnostic," *Physical Review Special Topics: Physics Education Research* 11, no. 2 (2015): 020130.
7. Smith et al., "Classroom Observation Protocol"; Lane and Harris, "New Tool."
8. The online course materials management system can be found at www.sei.ubc.ca.

9. For the use of clickers, see http://STEMclickers.colorado.edu; for the use of learning goals, see www.cwsei.ubc.ca/resources/learn_goals.htm; for SEI videos, see www.cwsei.ubc.ca/resources/SEI_video.html.
10. EOAS-SEI Times newsletter accessible at www.eoas.ubc.ca/research/cwsei/eossei-times.html.
11. See also www.eoas.ubc.ca/research/cwsei/courses.html.
12. Roger Fisher and William Ury, *Getting to Yes* (New York: Simon and Schuster Sound Ideas, 1987); John D. Bransford, Ann L. Brown, and Rodney R. Cocking, *How People Learn: Brain, Mind, Experience, and School* (Washington, DC: National Academies Press, 1999); Susan A. Ambrose, Michael W. Bridges, Michele DiPietro, Marsha C. Lovett, and Marie K. Norman, *How Learning Works: Seven Research-Based Principles for Smart Teaching* (Hoboken, NJ: John Wiley and Sons, 2010).
13. The training program schedule and materials are provided at www.cwsei.ubc.ca/resources/STLF-develop.htm.

5. What Was Achieved and What We Learned

1. Carl Wieman, "A Better Way to Evaluate Undergraduate Teaching," *Change: The Magazine of Higher Learning* 47, no. 1 (2015): 6–15.
2. A copy of the midway survey can be found at www.colorado.edu/sei/surveys/Sp10/SEI-FacultySurvey-Feb2010-PHYS.html.
3. Carl Wieman and Sarah Gilbert, "The Teaching Practices Inventory: A New Tool for Characterizing College and University Teaching in Mathematics and Science," *CBE-Life Sciences Education* 13, no. 3 (2014): 552–569.
4. Carl Wieman and Ashley Welsh, "The Connection between Teaching Methods and Attribution Errors," *Educational Psychology Review* 28 (2016): 645–648.
5. Stephanie V. Chasteen, Bethany Wilcox, Marcos D. Caballero, Katherine K. Perkins, Steven J. Pollock, and Carl E. Wieman, "Educational Transformation in Upper-Division Physics: The Science Education Initiative Model, Outcomes, and Lessons Learned," *Physical Review Special Topics: Physical Education Research* 11 (2015): 020110.
6. The 2012 and 2013 updates can be found at www.cwsei.ubc.ca/departments/earth-ocean.htm.
7. Items listed in this table are expanded upon in the agreement document, accessible at www.cwsei.ubc.ca/resources/files/Course_Transform_Expectations.pdf.
8. To see more details on this course, consult its webpage at www.cwsei.ubc.ca/departments/earth-ocean_TA.htm.

9. Previous issues of the newsletter can be accessed at www.eoas.ubc.ca/research/cwsei/eossei-times.html.
10. See the webpage at www.eoas.ubc.ca/research/cwsei.
11. This video series can be accessed at http://blogs.ubc.ca/wpvc.
12. Wieman and Gilbert, "Teaching Practices Inventory."
13. For more information on the survey and its findings, see Carl Wieman, Louis Deslauriers, and Brett Gilley, "Use of Research-Based Instructional Strategies: How to Avoid Faculty Quitting," *Physical Review Special Topics: Physics Education Research* 9, no. 2 (2013): 023102.
14. Chasteen et al., "Educational Transformation."
15. For more on the University of Colorado Learning Assistant Program, see https://laprogram.colorado.edu/.

6. The Post-Mortem: What Worked, What Didn't, and Why

1. John P. Kotter, *Leading Change* (Boston: Harvard Business Review Press, 1996).
2. Carl Wieman, "A Better Way to Evaluate Undergraduate Teaching," *Change: The Magazine of Higher Learning* 47, no. 1 (2015): 6–15.
3. For a thorough introduction to this concept, see Leon Festinger, *A Theory of Cognitive Dissonance*, vol. 2 (Palo Alto, CA: Stanford University Press, 1962).
4. A notable example of a study discussing this problem is Charles Henderson and Melissa H. Dancy, "Increasing the Impact and Diffusion of STEM Education Innovations," Invited paper for the National Academy of Engineering, Center for the Advancement of Engineering Education Forum, Impact and Diffusion of Transformative Engineering Education Innovations, 2011.

Appendix 1. SEI Course Transformation Guide

1. G. Gibbs and C. Simpson, "Conditions under Which Assessment Supports Student Learning," *Learning and Teaching in Higher Education* 1 (2004): 3–31.
2. Effective techniques involve designing assignments to be of obvious benefit to the learning of the student; they should have substantial overlap with the exams and have some portions of the assignment that involve "explaining in your own words."
3. S. Bonham, "Reliability, Compliance, and Security in Web-Based Course Assessments," *Physical Review Special Topics: Physics Education Research* 4 (2008): 010106.

4. C. Crouch and E. Mazur, "Peer Instruction: Ten Years of Experience and Results," *American Journal of Physics* 69 (2001): 970–977.
5. See "Just-in-Time Teaching," Department of Physics, Indiana University—Purdue University Indianapolis, 2006, http://jittdl.physics.iupui.edu/jitt.
6. K. Topping, "Peer Assessment between Students in Colleges and Universities," *Review of Educational Research* 68, no. 3 (1998): 249–276.
7. B. Gilley and B. Clarkston, "Collaborative Testing: Evidence of Learning in a Controlled In-Class Study of Undergraduate Students," *Journal of College Science Teaching* 43, no. 3 (2014): 83–91; G. Rieger and C. Heiner, "Examinations That Support Collaborative Learning: The Students' Perspective," *Journal of College Science Teaching* 43, no. 4 (2014): 41–47.

ACKNOWLEDGMENTS

The Science Education Initiative and this book would never have been possible without the help of many people. The most important of them is my wife, Sarah Gilbert, whose encouragement and hard work first made the SEI a reality, and then the UBC SEI a success. She also helped a great deal in the writing of this book. Stephanie Chasteen contributed significantly, too, particularly in the writing of Chapter 4 on the science education specialists and with the collection and analysis of data from the CU SEI. Kathy Perkins played an invaluable part in first defining what an SES could and should be, and then managing the CU SEI. This book would never have been written without encouragement and helpful advice from Peter Lepage.

It is impossible to list the hundreds of thoughtful administrators and faculty members at CU and UBC whose contributions made the SEI possible. Finally, I would like to acknowledge the dozens of dedicated and extremely hardworking SESs who were the heart of the SEI and from whom I learned an enormous amount.

INDEX